MEMS Reference Shelf

Series Editors:

Stephen D. Senturia
Professor of Electrical Engineering, Emeritus, Massachusetts Institute
of Technology, Cambridge, Massachusetts

Roger T. Howe
Professor, Department of Electrical Engineering, Stanford University,
Stanford, California

Antonio J. Ricco
Small Satellite Division, NASA Ames Research Center, Moffett Field, California

For further volumes:
http://www.springer.com/series/7724

Allyson L. Hartzell · Mark G. da Silva ·
Herbert R. Shea

MEMS Reliability

Foreword by Stephen D. Senturia

 Springer

Allyson L. Hartzell
Lilliputian Systems, Inc.
36 Jonspin Road
Wilmington, MA 01887-1091, USA
ahartzell@lsinc.biz

Mark G. da Silva
Analog Devices Inc.
RSTC, MS-112
804 Woburn Street
Wilmington, MA 01887, USA
mark.dasilva@analog.com

Herbert R. Shea
Microsystems for Space
   Technologies Laboratory
EPFL
Rue Jaquet-Droz 1
CH-2002, Neuchatel, Switzerland
herbert.shea@epfl.ch

ISSN 1936-4407
ISBN 978-1-4419-6017-7   e-ISBN 978-1-4419-6018-4
DOI 10.1007/978-1-4419-6018-4
Springer New York Dordrecht Heidelberg London

Library of Congress Control Number: 2010935580

Printed on acid-free paper

Springer is part of Springer Science+Business Media (www.springer.com)

Foreword

How long will this MEMS device last? An important question. In fact, for any product that is to succeed in the marketplace, an essential question. The answer, of course, can only be provided on a statistical basis, and the data on which that statistical answer is based come from measurements of in-service lifetimes under aggressive conditions of temperature, humidity, chemical exposure, shock, or other challenges. The design of these accelerated lifetime tests, as well as the corresponding interpretation of data from them, depends on a deep knowledge of the device structure, the constituent materials, the physical mechanisms that might lead to device failure, the dependence of failure rates on such factors as temperature, and both physical and statistical modeling of failure modes. The knowledge base on which this life-testing depends comes from experience with individual device failures, whether so-called infant mortality due to manufacturing defects, or well-understood mechanisms, such as corrosion or metal fatigue, that only appear after long exposure to appropriate challenges. Developing that knowledge base, and using it wisely, is a tall order.

Ultimately, achieving a high level of device reliability requires a disciplined approach to device design, manufacturing, and quality control. If one can anticipate failure modes and thus design devices so as to minimize the risk of failure, if one can rigorously monitor manufacturing procedures to assure that completed devices are free of flaws, if one can design accelerated life tests based on sound device physics and associated models, and if one can implement the discipline to check every device failure, document every manufacturing step, and trace back to identify the root cause of every observed failure, then one has a hope of developing reliable products with predictable in-use lifetimes.

This volume is the ideal starting place for anyone seeking to address these issues of MEMS reliability. Even more than the field of MEMS itself, MEMS reliability touches everything: from basic chemistry and physics to statistical methods of lifetime prediction, even to organizational issues needed to address device reliability at every step of device design and development. I am pleased to present it to the MEMS world as a key component of the Springer MEMS Reference Shelf.

Editor-in-Chief   Stephen D. Senturia
Springer MEMS Reference Shelf
Brookline, MA
June 2010

Preface

The widespread growth and acceptance of microsystem technology in diverse applications from consumer electronics to space and military hinges on products achieving a suitable balance of quality and cost. Quality essentially implies that a product performs as specified in the datasheet, which essentially means that it performs reliably. The fundamental approach to MEMS device reliability employs some of the same basic concepts and methodologies established in high volume automotive and IC manufacturing; including FMEA (failure mode and effects analysis – root cause), DfM (Design for Manufacturability), DfR (design-for-reliability) and lifetime prediction. A major challenge in MEMS is the shear diversity of potential applications, novel materials and processes, unique sensing and actuation principles, and manufacturing techniques, and hence the focus of this book is on reliability techniques and methodologies as applied to MEMS devices.

MEMS Reliability, especially the study of reliability physics, is a vast area that is still in its infancy in academic coursework. University research, government laboratory research, and consortia studies have been and continue to contribute invaluable advances in MEMS reliability physics. However, working in industry and mass producing hundreds of millions of reliable MEMS devices, some of which are intended for safety critical applications, provides a very different perspective. The authors of this textbook all have multiple years of academic and industry experience in MEMS design, fabrication, production, and reliability, and each have their own areas of expertise that have been brought together to produce a book that is scientific in its approach and coherent in its structure, with topics from all worlds of MEMS reliability study as well as case studies of successful product reliability development. Our hope is that this text will serve as a useful guide for setting up a reliability programs for real-world products and to spur further interest in solving some of the fundamentally challenging problems in the field.

This is not an edited book, and is therefore unique in MEMS Reliability texts because the book can be used by academia in preparing the student for industry work, and by industry engineers as a reference guide for the reliable manufacture of MEMS in any volume. Bringing together reliability statistics, acceleration testing, manufacturing failure modes, design for reliability, in-use physics of failure, root cause analysis, failure analysis, testing methods for MEMS, qualifications of MEMS, and continuous improvement methodologies was key to presenting this challenging subject in a synergistic manner.

Wilmington, Massachusetts   Allyson L. Hartzell
Wilmington, Massachusetts   Mark G. da Silva
Neuchatel, Switzerland   Herbert R. Shea

Acknowledgements

Allyson Hartzell is thankful for the experience of working at Analog Devices Incorporated Micromachined Products Division (MPD), and would particularly like to thank Bill O'Mara, Ira Moskowitz, Ray Stata, and the late Bob Sulouff, for their support during this time. Ms. Hartzell first started working on MEMS Reliability in 1997 at ADI; the Analog Devices examples in this book were published while she was working in the MPD division. Ms. Hartzell would also like to thank Paul Bierden of Boston Micromachines Corporation (BMC) as well as Dr. Jason Stewart of BMC and Professor Tom Bifano of Boston University for their reviews and comments. Allyson Hartzell is thankful to David for his unwavering support during the writing of this book, and to my mother Silvana and my family who have always believed in me.

Herb Shea entered the world of MEMS reliability when he joined the Bell Labs MEMS reliability group. He very warmly thanks the entire Bell Labs LambdaRouter™ team, and in particular Dr. S. Arney, Dr. A. Gasparyan, Dr. M. E. Simon and Dr. F. Pardo. Many of the examples in Chapter 4 are based on the work of the Bell Labs MEMS team. HS thanks Subramanian Sundaram who developed the shock and vibration response models presented in Chapter 4 while an intern at the EPFL. HS is very grateful to Véronique for her patience and constant support and to my parents for their love and selfless dedication.

Mark da Silva is thankful to all his current and former colleagues at Coventor, Exponent and Analog Devices, as well as numerous other colleagues in the MEMS industry for their contributions, teachings and discussions over the years. Without their help Mark da Silva would not have such an appreciation and understanding of the field of MEMS reliability. Mark would also like to personally thank Mike Judy, Shawn Cunningham, Robert Giasolli, and Stuart Brown for their assistance in putting together the ASME MEMS Reliability short course which provided an invaluable experience and helped comprehend the knowledge gaps in MEMS reliability. Additionally, Mark is grateful to Stephen Senturia, Roger Howe, Stephen Bart and Steven Elliot (Springer) for their reviews and valuable suggestions. Lastly, no acknowledgement from Mark would be complete with thanking his family–his wife Anna, his son Max, his parents and brother for their patience, love and constant encouragement during this effort.

Contents

Chapter 1
Introduction: Reliability of MEMS

The development of Micro-Electro Mechanical Systems (MEMS) and introduction of MEMS-enabled products in the market have made amazing strides in the last two decades; fulfilling a vision of "cheap complex devices of great reliability".[1] MEMS are integrated micro-scale systems combining electrical, mechanical or other (magnetic, fluidic/thermal/etc.) elements typically fabricated using conventional semiconductor batch processing techniques that range in size from several nanometers to microns or even millimeters [1]. These systems are designed to interact with the external environment either in a sensing or actuation mode to generate state information or control it at a different scale.

In recent years, MEMS technology has gained wide-spread acceptance in several industrial segments including automotive, industrial, medical and even military applications. The size and growth of the MEMS market is typically represented in volume of a particular kind of sensor device, and in 2009 this market was roughly US$7 Billion and was dominated by pressure sensors, accelerometers, optical devices and microfluidic devices (Fig. 1.1 below)[2] and represents roughly 8–10 billion units.

MEMS present several daunting technical challenges quite unlike those seen in typical semiconductor microelectronics which have no moving parts. In comparison, MEMS designers create a variety of different 3D structures and highly complex shapes (see Fig. 1.2 below) all at similar micrometer scales using a variety of materials. Another unique challenge in MEMS is that the end-product functionality is often tightly linked to the process used to create it leading to the "one product, one process." This is in marked contrast to the IC industry where many products share a common process (i.e., there is no equivalent of a "32 nm node" for MEMS).

MEMS process or fabrication technology has made great strides in recent years to mass fabricate at the micro-scale with a variety of materials (besides conventional semiconductor materials) using standard photolithographic processes for high volume MEMS device fabrication.

---

[1] "The world has arrived at an age of cheap complex devices of great reliability, and something is bound to come of it" – Dr. V. Bush (1945) [2].

[2] Databeans estimate.

A.L. Hartzell et al., *MEMS Reliability*, MEMS Reference Shelf,
DOI 10.1007/978-1-4419-6018-4_1, © Springer Science+Business Media, LLC 2011

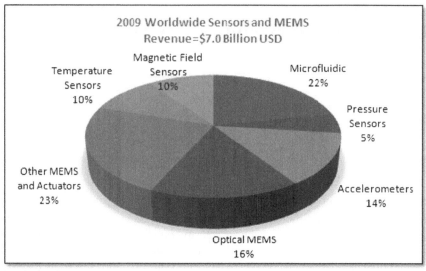

2009 Worldwide Sensors and MEMS
Revenue=$7.0 Billion USD

Magnetic Field
Temperature   Sensors   Microfluidic
Sensors   10%   22%
10%
                                                              Pressure
                                                              Sensors
                                                               5%
Other MEMS
and Actuators   Accelerometers
23%   14%

Optical MEMS
16%

Fig. 1.1 Estimate of World Wide MEMS Market 2009 (reprinted with permission Copyright – Databeans)

Fig. 1.2 Polysilicon fabricated accelerometer (reprinted with permission Copyright – Analog Devices)

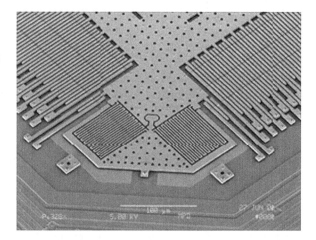

The road has not always been smooth, and evolution of MEMS fabrication technology has taken the better part of the last two decades. In this time, MEMS based products have crossed the threshold of prototype volumes into large-scale volume production. Examples such as the Freescale MPX Series pressure sensor, the Analog Devices *ADXL series* accelerometers, and the Texas Instruments *DLP*® mirror are but a few products that have successfully achieved the performance and cost targets necessary to displace competing technologies in specific markets (Fig. 1.3).

Period	Applications	Impact on the MEMS Industry	Major Players in Production
1990 -	▼MEMS built into equipment (inkjet heads, etc.)	▼Establish mass-production	Equipment manufacturers Automobile manufacturers
2000 -	▼MEMS devices (accelerometers, etc.)	▼Accelerate low-cost technologies and expand production infrastructure	Semiconductor manufacturers Parts
2005 -	▼Integrated MEMS (devices merging CMOS and LSI, etc.)	▼Develop standards for production processes and advance integration with LSI products	Semiconductor manufacturers MEMS foundries
2010 -	▼MEMS integrated devices from semiconductor manufacturers	▼Develop an IP core for MEMS functions	Silicon/MEMS foundries Semiconductor manufacturers

Fig. 1.3 MEMS development history (reprinted with permission Copyright – MMC)

On average, each of these product development efforts lasted several years from initial concept to final volume production and market insertion, although improvements in time-to-market have been observed with successive generations of products. By far the most significant time-consuming factor in each case has been the persistence of a "traditional" manufacturing approach (Fig. 1.4 below), where the engineering of the product for volume manufacturing has gone through many cycles of learning and consequently has taken much longer than anticipated. The novelty of MEMS technology, lack of adequate design tools, a lack of "standard" process flows, the complex interaction of packaging and MEMS device, and MEMS reliability, are challenges that have hindered quicker time to market. Although DFM (Design for Manufacturing) and TQM (Total Quality management) strategies exist in almost all industries today [3], the adoption of a comprehensive design methodology that links all product engineering groups in the MEMS industry was lacking in the early days [4]. In recent years, product development methodologies for MEMS product design are grounded in powerful top-down design tools. Today, concurrent engineering practices (Fig. 1.5) have reaped benefits in terms of *faster design cycles and a faster path to volume manufacture*. A major challenge continues to be the reliability of the MEMS enabled product in the intended application.

This book was written to aid in the improvement of MEMS product reliability by providing an understanding of the science, and best practices, as well as to document

Fig. 1.4 Traditional MEMS product development cycle (reprinted with permission Copyright – Sensors Expo)

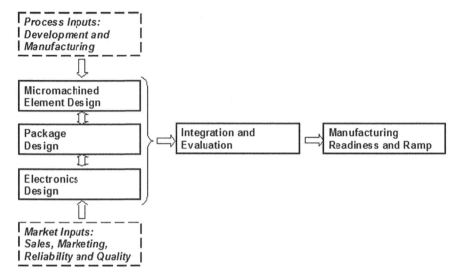

Fig. 1.5 Product development flow based on concurrent engineering practices (reprinted with permission Copyright – Sensors Expo)

the methodology to drive improvement within a MEMS enabled product. The intent is to provide readers with a valuable reference containing a detailed description of failure mechanisms, understanding of reliability physics, lifetime prediction, test methods, and numerous examples for making improvements to reliability in all types of MEMS products. With the MEMS industry as diverse as it is today, a reference of industry best-practices is helpful in providing new product development efforts with a guide to address specific reliability challenges ahead of time or through design, thereby reducing time-to-market. The following figure (Fig. 1.6) illustrates the linkage between the topics covered in this book, and although obviously simplified, yet should allow the reader to understand the connections between the topics of MEMS reliability.

This book begins with this introductory chapter on the need for improved understanding of MEMS reliability issues.

Chapter 2 provides a review of reliability statistics for lifetime prediction, and includes the Weibull, Lognormal and Exponential distributions. The *bathtub* curve concept is presented as well as acceleration factors of physics of failure, and accelerated testing in MEMS. This chapter also introduces the reader to MEMS reliability through three basic case studies. Acceleration of hinge related creep failure in the DLP® mirror from Texas Instruments and a predictive model of mechanical shock related to stiction in an accelerometer test vehicle from Analog Devices are presented. These are examples of two very different high-volume and high reliability MEMS products. The third case study is a MEMS product with great potential that has not yet entered the marketplace due to reliability challenges; MEMTronics' RF MEMS product is reviewed.

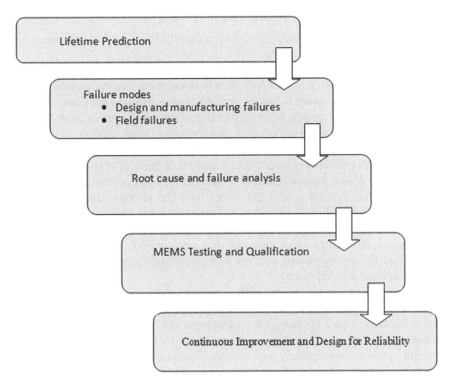

Fig. 1.6 Reliability topics

Chapter 3 examines failure mechanisms and modes introduced in MEMS technology and products during the design and manufacturing stages. This chapter
examines the impact of design and manufacturing failure modes that have their origin in the product development phase on the reliability of the final product. These
failures include both functionally analyzed and non-analyzed behaviors that are
repeatedly seen to be major factors in the life of the product. The second half of
the chapter presents manufacturing (or process) failures due to typical MEMS fabrication process failure modes including defects like contamination, release stiction,
intrinsic and extrinsic material failures, handling and packaging failure modes.

Chapter 4 covers the physics of failures modes in the field (operational and non-
operational). This comprehensive chapter covers mechanical mechanisms, electrical
mechanisms and environmental mechanisms by combining theory and data. For
each failure mode, mitigation techniques and performance/reliability trade-offs are
presented. It should be kept in mind that each failure mechanism will have a different level of predominance depending on the device, fabrication process, and the
operational environment and range of the device. Mechanical failure modes specific to MEMS are discussed in detail, including fracture, shock resistance, fatigue,
and plastic deformation. Electrical failure modes include dielectric breakdown,

dielectric charging, ESD, and electromigration. Environmental failures cover the effects of ionizing and non-ionizing radiation on MEMS, as well as different types of corrosion for metals and silicon.

Chapter 5 defines strategies for identifying root cause and begins with the Failure Modes and Effect Analysis (FMEA), which is an excellent tool to determining root cause and corrective action to assure that failure mechanisms are contained and eliminated. An optical switch Reliability FMEA is presented as an example to teach the reader how to use this methodology. A substantial failure analysis section that describes popular failure analysis techniques with MEMS-based analytical results is also included. A reliability program must contain strategies for identifying potential failure modes, failure mechanisms, risk areas in design and process, and corrective action strategies. Containment of the failure is of crucial importance to minimizing or mitigating the field failure rate while a proper root cause is identified, and corrective action is developed and finally implemented into production. For MEMS technologies, the use of proven methodologies (such as FMEA and failure analysis techniques) to identify potential failure modes and mechanisms are an important part of the reliability approach.

Chapter 6 contains testing and qualification processes and procedures used in MEMS, and are presented with discussions of relevant reliability. This chapter introduces for the first time the unique test equipment and reliability test methods used in the MEMS industry along with quality standards for various target industries that include automotive and military applications. Examples include test data for MEMS specific test equipment as well as qualification and reliability studies.

Chapter 7 offers a summary of the best practices to improve reliability in a MEMS product. The information in this chapter is a synopsis of much more in-depth work and readers should reference other sections of this book or articles listed at the end of this chapter for more details. The chapter discusses the yield-reliability connection specifically to MEMS products where it is common to screen parts at final test for various weaknesses that could result in potential field failures, including ones that impact life of the part. The importance of process and material property characterization [6] is discussed next as well as the use of common test structures and Process Control Monitors. In typical CMOS processes, the importance of process stability and reproducibility is quite well recognized but in MEMS the significance takes on a new dimension because of the additional specialized process steps. Common techniques for yield and quality enhancements are discussed in some depth, as the yield/reliability link has to be kept in mind at all times. Finally, the topic of Design-for-Reliability (DfR) addresses the topic of design methodology that considers probable failure of the device as part of the design process.

In summary, as this unique text was written by authors with extensive industry reliability experience, we have distilled the best practices from a variety of MEMS product development efforts to provide the reader with a clear methodology for developing a solid and comprehensive MEMS reliability program.

References

1. Senturia, S.D. (2001) *Microsystem Design*. Dordrecht: Kluwer.
2. Bush, V. (July 1945) As We May Think, *Atlantic Magazine* (http://www.theatlantic.com/magazine/archive/1969/12/as-we-may-think/3881/)
3. Bralla, J.G. (1998) *Design For Excellence*. London, UK: McGraw-Hill Book Co.
4. da Silva, M.G., Giasolli, R., Cunningham, S., DeRoo, D. (2002) MEMS design for manufacturability. *Sensors Expo*. Boston.
5. ASME (2003) *Course 469: MEMS Reliability Short Course*. New York: ASME.
6. (2003). *INTEGRRAM Metal-Nitride Prototyping Kit – Design Handbook, Metal-Nitride Surface Micromachining*. QinetiQ Ltd. & Coventor Sarl.

Chapter 2
Lifetime Prediction

2.1 Introduction

Reliability continues to be one of the critical drivers for MEMS acceptance and growth. Emerging technologies require marketplace acceptance in order to be designed into high volume and critical applications. Thus, the field of reliability physics must be approached at the most fundamental level when evaluating and predicting micromachined product field performance over the lifetime of the product. The lifetime prediction portion of the reliability program is seen in Fig. 2.1.

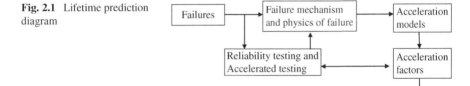

Fig. 2.1 Lifetime prediction diagram

Reliability testing is required to accelerate the lifetime of the MEMS part using acceleration factors, for proper lifetime prediction. This chapter will cover basic reliability statistics and failure distributions used in lifetime prediction.

Development of acceleration factors and reliability testing will also be covered. Case studies for two successful MEMS products, Texas Instruments' digital micro-mirror device (DLP®) and Analog Devices' accelerometer, include physics of failure, reliability testing and statistical field predictions. A third case study is a product that has yet to be put into volume production: RF MEMS.

2.2 Mathematical Measures of Reliability

This section will cover the most popular mathematical statistics used in reliability. The survivor or reliability function, cumulative distribution function, probability

A.L. Hartzell et al., *MEMS Reliability*, MEMS Reference Shelf,
DOI 10.1007/978-1-4419-6018-4_2, © Springer Science+Business Media, LLC 2011

distribution function, hazard function, and the bathtub curve concept are included and related. These functions are used to measure failure distributions and predict reliability lifetimes. The Exponential, Weibull and Lognormal distributions will be covered.

Reliability is the probability of the product performing properly under typical operating conditions for the expected lifetime intended, and an expression to define reliability is:

$$R(t) = 1 - F(t) \tag{2.1}$$

Here, $R(t)$ is the reliability function, also called the survivor function. This is defined as the probability of operating without failure to time t. $F(t)$ is the cumulative failure distribution function (CDF). In reliability, $F(t)$ is the probability that a randomly chosen part will fail by time t. A lifetime distribution model $f(t)$ is the probability density function (PDF) over the time range 0 to ∞ (infinity). The relationship between the CDF and PDF is shown in (2.2) and (2.3).

$$F(t) = \int_0^t f(t')dt' \tag{2.2}$$

$$f(t) = \frac{d}{dt}F(t) \tag{2.3}$$

The hazard rate $h(t)$ is also known as the instantaneous failure rate. This is the probability that failure will occur in the next time interval divided by the reliability $R(t)$ (the probability of operating without failure up to that time interval) [1].

$$h(t) = \frac{f(t)}{1 - F(t)} = \frac{f(t)}{R(t)} \tag{2.4}$$

This can also be written as

$$h(t) = -\frac{1}{R(t)}\frac{dR(t)}{dt} \tag{2.5}$$

Which is equivalent to

$$h(t) = -\frac{d}{dt}(\ln R(t)) \tag{2.6}$$

The integral of the hazard rate is the cumulative failure rate (cumulative hazard rate)

$$H(t) = \int_0^t h(t')dt = -\ln R(t) \tag{2.7}$$

The hazard rate $h(t)$ or instantaneous failure rate has dimensions of (time^{-1}). Since $R(0) = 1$ (no failures at time zero), the reliability rate over a time period t is the exponential of the cumulative hazard rate in that same time period t.

$$R(t) = e^{-\int_0^t h(t')\,dt'} \tag{2.8}$$

An important quantitative reliability concept is how long the population will survive without a failure. This is also termed mean time to failure (MTTF), more specifically, the mean-time to the first failure [2].

$$\text{MTTF} = \bar{t} \equiv \int_0^\infty tf(t)\,dt \tag{2.9}$$

2.3 Reliability Distributions

2.3.1 Bathtub Curve

The distribution of failures over the lifetime of the product population is critically important to the MEMS reliability physicist. Using these concepts, distribution functions can be developed and used for predictive purposes. A hazard rate that changes over the lifetime of the product, starting high, reducing, and increasing towards the end of the product life, is also termed the "bathtub curve" (Fig. 2.2). The population will have defective items that will fail within the first few weeks to months of the product lifetime (infant mortality) is termed the bathtub curve because of the shape of the curve itself. An ideal failure behavior is to eliminate the failures due to defects in the infant mortality portion of the curve through burn-in and/or defect reduction programs, and to not operate the product into the wear-out phase. The operational life is within the typically constant hazard rate section of the curve.

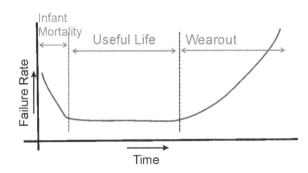

Fig. 2.2 The bathtub curve, showing three stages over the device lifetime: high initial failure due to infant mortality, constant failure rate over the useful lifetime, and increased failure rate as the devices age

Fig. 2.3 Illustration of how the bathtub curve can be viewed as the sum of the three failure rates. Reprinted with permission. Copyright 1993 Springer Business and Media [2]

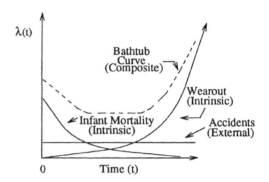

Figure 2.3 illustrates how the bathtub curve can be the composite of three failure rates: infant mortality, wear out, and externally induced failure.

For proper lifetime distribution modeling, individual failure mechanisms must be modeled independently, and there must be only one population. If there are multiple populations (or subpopulations) within the data, they must be individually extracted and statistically analyzed as single populations. There are various time to failure distributions to express population lifetime behavior statistically. Three popular statistical reliability distributions are the Exponential, the Weibull and the Lognormal.

2.3.2 Exponential Distribution

The exponential distribution is the least complex of all lifetime distribution models. The failure rate or hazard rate, $h(t)$, is λ. The failure rate is a constant in this model, which is suitable for the stable failure rate regime in Fig. 2.2, the bathtub curve. The reliability (2.10), the cumulative distribution function (CDF, (2.11)) and the probability distribution function (PDF, (2.12)) are shown below (Figs. 2.4, 2.5 and 2.6).

$$R(t) = e^{-\lambda t} \tag{2.10}$$

$$F(t) = 1 - e^{-\lambda t} \tag{2.11}$$

$$f(t) = \lambda e^{-\lambda t} \tag{2.12}$$

The mean time to failure of the exponential function is simply the inverse of the failure rate λ.

$$\text{MTTF} = 1/\lambda \tag{2.13}$$

Fig. 2.4 Cumulative distribution function $F(t)$ for exponential distribution

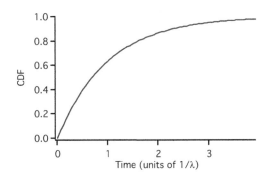

Fig. 2.5 PDF for exponential distribution

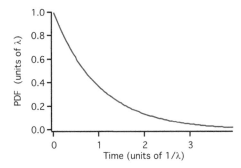

Fig. 2.6 Exponential distribution hazard rate

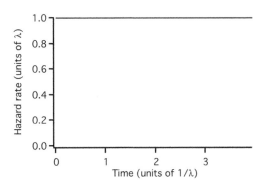

2.3.3 Weibull Distribution

The Weibull distribution function is used to fit various shapes of reliability curves. The Weibull function can be expressed in multiple ways [3]. The Weibull distribution expression below is the probability of survival $R(t)$ between time zero and time t [4].

$$R(t) = e^{-\left(\frac{t-\gamma}{\alpha}\right)^{\beta}} \tag{2.14}$$

There are three Weibull reliability curve fit parameters in even the basic form of the Weibull function. They are (1) β, the shape parameter, (2) γ, the location parameter (also known as the defect initiation time parameter), and (3) α, the characteristic life or scale parameter. This Weibull distribution function can have two variants: the two-parameter distribution and the three-parameter distribution. The difference between the two variants is whether or not failures start at time zero. If failures do start at time zero, the defect initiation time parameter (also known as location parameter) is zero and the Weibull exponential expression is reduced to

$$R(t) = e^{-(\frac{t}{\alpha})^{\beta}} = f(t)/h(t) \tag{2.15}$$

When $\beta = 1$, equation (2.15) becomes the exponential model(2.10), with $\alpha = 1/\lambda$, the MTTF (2.13). The two parameter fit model is commonly used in reliability life predictions. The PDF of the two parameter Weibull model is in Fig. 2.7

$$f(t) = \frac{\beta}{t} \left(\frac{t}{\alpha} \right)^{\beta} e^{-(\frac{t}{\alpha})^{\beta}} \tag{2.16}$$

The CDF of the two parameter Weibull model is in Fig. 2.8 while Hazard function is in Fig. 2.9

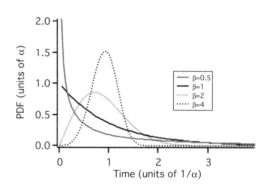

Fig. 2.7 Weibull function PDF in units of α, varying β

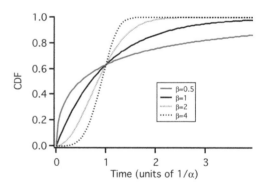

Fig. 2.8 CDF of Weibull function, varying β

$$F(t) = 1 - e^{-\left(\frac{t}{\alpha}\right)^{\beta}} \tag{2.17}$$

The cumulative failure rate of the two parameter Weibull model (cumulative hazard rate) is expressed as

$$H(t) = \left(\frac{t}{\alpha}\right)^{\beta} \tag{2.18}$$

The instantaneous failure rate is

$$h(t) = \frac{\beta}{\alpha}\left(\frac{t}{\alpha}\right)^{\beta-1} \tag{2.19}$$

Sandia has published a good example of Weibull failure data on their MEMS microengine [5, 6]. In this study, 41 microengines (Figs. 2.10 and 2.11) were driven to failure with the SHiMMer test platform (Chapter 6), and the data was fit to both Weibull and lognormal functions. The cumulative failure rate is plotted as a function of accumulated cycles. A production-ready process will have a β value of 0.5 to 5 as evaluated by the Weibull function. The data in Fig. 2.11 has a β of 0.22, indicating that the data is widely dispersed. This plot shows 50% failure at 10^7 cycles.

Fig. 2.9 Hazard rate for Weibull function

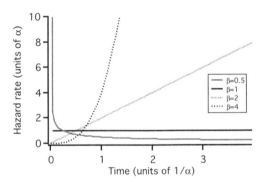

2.3.4 Lognormal distribution

The other popular reliability statistical distribution, the lognormal time to failure distribution, is as it is named, log normally distributed. The lognormal (also called Gaussian) distribution PDF is (Figs. 2.12 and 2.13)

$$f(t) = \frac{1}{\sigma t \sqrt{2\pi}} e^{\left(-\frac{(\ln(t) - \ln(T_{50}))^2}{2\sigma^2}\right)} \tag{2.20}$$

Fig. 2.10 Gear structure of
Sandia's microengine, made
with SUMMiT^{TM} process.
Courtesy of Sandia National
Laboratories, SUMMiT(TM)
Technologies,
www.mems.sandia.gov [6]

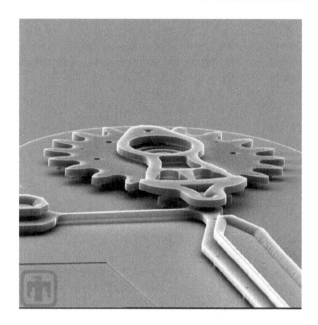

Fig. 2.11 Weibull probability
plot of microengine failures.
Courtesy of Sandia National
Laboratories, SUMMiT(TM)
Technologies,
www.mems.sandia.gov [5]

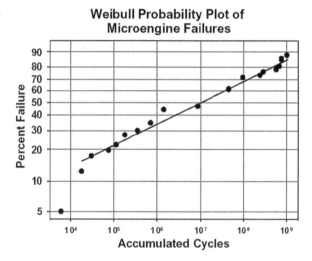

The cumulative distribution function $F(t)$ is below in (2.21), while the solution is
in (2.22).

$$F(t) = \int_0^T \frac{1}{\sigma t \sqrt{2\pi}} e^{\left(-\frac{(\ln(t)-\ln(T_{50}))^2}{2\sigma^2}\right)} dt \qquad (2.21)$$

Fig. 2.12 PDF lognormal function

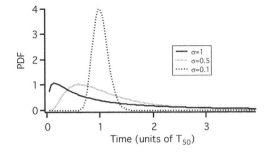

Fig. 2.13 CDF lognormal distribution

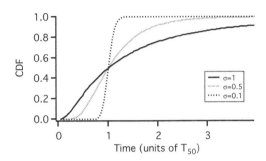

$$F(t) = \Phi \left[\frac{\ln(t/\tau)}{\sigma} \right] \tag{2.22}$$

where $\Phi(z) = \frac{1}{2} \left[1 + Erf \left(z/\sqrt{2} \right) \right]$

The shape parameter sigma σ (standard deviation) is the slope of the time to failure vs. the cumulative percent failure on a log scale. The remaining functions can be calculated using the equations in Section 2.3 (Fig. 2.14).

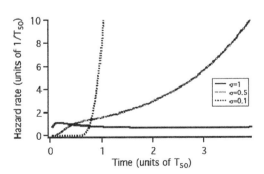

Fig. 2.14 Hazard rate for lognormal distribution

Fig. 2.15 Lognormal
probability plot of sandia
SUMMiT^{TM} microengine
failure. Courtesy of Sandia
National Laboratories,
SUMMiT(TM) Technologies,
www.mems.sandia.gov [5]

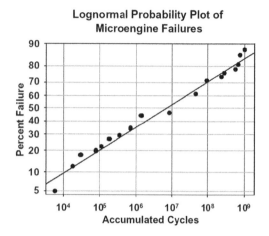

Figure 2.11 is the Sandia microengine failure data analyzed with the Weibull
distribution function. This same data is analyzed with the lognormal function in
Fig. 2.15.

In this analysis, $\sigma = 5$. This indicates a large spread in the lifetime. The range
typical for semiconductor products is 0.1–1. The time to 50% failure is 7.8 million
cycles when this data is analyzed via the lognormal distribution.

Any analysis must be studied for bimodal distributions. This lognormal example
can be also analyzed as two separate populations in Fig. 2.16. The early life failure
sub-population (infant mortality in Fig. 2.2) has a time to 50% failure of 140,000
cycles while the wear-out portion of the curve has a time to 50% failure of 250

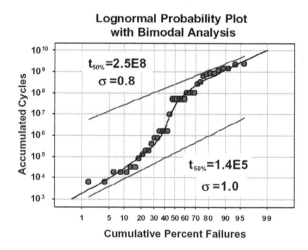

Fig. 2.16 Bimodal analysis
using lognormal fit. Courtesy
of Sandia National
Laboratories, SUMMiT(TM)
Technologies,
www.mems.sandia.gov [5]

million cycles. Both of these fits have acceptable sigmas, which indicate that this analysis of the data is most accurate versus treatment as a single population.

Most failure mechanisms are treated with the lognormal distribution, but some, including solder fatigue, must be modeled with the Weibull distribution. Thus, for new physics of failure in MEMS, it is recommended to use the Weibull distribution unless it is proven that the lognormal distribution will accurately fit the shape of the empirical failure distribution.

2.3.5 Acceleration Factors

Reliability testing at accelerated conditions is critical to generating lifetime data in a much shorter period of time. Release of a reliable product to market is dependent on this concept. Stresses (examples are elevated temperature, temperature cycling, applied voltage, and relative humidity) experienced in the use environment are "accelerated", or increased to a level to accelerate the time to failure of an individual failure mechanism. The key is to create the same failure mechanism as occurs in use conditions. Development of an acceleration model is performed through knowledge of the physics of failure. An acceleration factor is calculated as compared to the use conditions. A summary table of some known MEMS failure mechanisms and accelerating stresses is below [7] in Table 2.1. Chapters 3 and 4 detail these failure mechanisms.

The field of MEMS does not have a long history of known failure models and easily obtained acceleration factors when compared to the more seasoned semiconductor industry. Standard integrated circuit reliability science has undergone many years of study to accurately predict lifetimes [8]. Table 2.2 is a chart of commonly used semiconductor packaging and assembly acceleration models [9].

There are examples in semiconductor physics of failure where many models exist for the same failure mechanism. Table 2.3 is a table that includes many existing corrosion models. The voltage term in these models must be developed as it is typically not known. In the case of established models, literature reviews are recommended to assure that the proper model is used.

To properly use acceleration models and compare to use conditions, operating environment, storage environment (non-operating) and the lifetime of the product must be known. Although this is unique to each product development effort and requires a discussion between the customer and supplier, Table 2.4 is a guideline [9] developed for the semiconductor community that can be applied to MEMS products. Examples of some major market segments are:

- Indoor: Computers, Laboratory test equipment, Projectors, Printers, etc.
- Consumer Portable: Cell phones, PDA's, Portable Laptop and Notebook PCs, etc.
- Other: Automotive, Outdoor telecommunications equipment,

Table 2.1 Examples of MEMS failure mechanisms and accelerating factors

Failure mechanism	Accelerating factors	Additional comments
Cyclic fatigue	No. of cycles, maximum applied strain, humidity	Models exist for this failure mechanism in mechanical engineering texts and literature, as well as some MEMS structures.
Creep (plastic deformation)	Temperature, applied strain	Well understood materials science field.
Stiction	Humidity, shock, vibration	Difficult to model. Surface conditions are critical.
Shorting and open circuits	Electric field, temperature, humidity	Well understood field, yet the geometries in MEMS and materials used could make this difficult to model for some structures. Again, processing effects can be critical.
Arcing	Electric field, gas pressure, gas composition	Small gaps are prone to this in specific environments. Breakdown voltage relationships should be investigated.
Dielectric charging	Electric field, temperature, radiation, humidity	Some MEMS structures such as RF MEMS are particularly susceptible to this.
Corrosion	Humidity, voltage, temperature	Polarity is important if accelerating anodic corrosion.
Fracture due to shock and vibration	Acceleration, frequency (resonance), vacuum	Models exist for this failure mechanism in mechanical engineering texts and literature, as well as some MEMS structures. Micro-scale materials properties are needed.

Not included here are extreme use environments such as space [10]. (A section on space radiation physics of failure is in Section (4.4.1).) Standards for space missions and other extreme environment MEMS applications exist, yet qualification testing for space is typically mission-specific. Standards for general qualification testing are covered in Chapter 6.

Table 2.2 Table of commonly used packaging and assembly acceleration models for semiconductors

Mechanism	Model	Assumptions
Temperature, humidity mechanisms	Peck's $TF = A_0 \times RH^{-N} \times \exp[E_a/kT]$ AF (ratio of TF values, use/stress) $AF = (RH_{stress}/RH_{use})^{-N} \times \exp[(E_a/k)(1/T_{use} - 1/T_{stress})]$ When calculating variables that stay constant between Stress 1 and Stress 2 they will drop out of the equation.	• AF = acceleration factor • TF = time to failure • A_0 = arbitrary scale factor • V = Bias voltage • RH = relative humidity as % • N = an arbitrarily determined constant • E_a = activation energy for the mechanism (0.75 eV is conservative) • k = Boltzmann's constant, 8.162×10^{-5} eV/°K • T = temperature in Kelvin. There are other models for THB mechanisms and they should be checked for the fit to the data.
Thermal effects	Arrhenius $TF = A_0 \times \exp[E_a/kT]$ AF (ratio of TF values, use/stress) $AF = \operatorname{Exp}[(E_a/k)(1/T_{use} - 1/T_{stress})]$	• AF = acceleration factor • TF = time to failure • A_0 = arbitrary scale factor • E_a = activation energy for the mechanism (0.75 eV is conservative) • k = Boltzmann's constant, 8.162×10^{-5} eV/°K • T = temperature in Kelvin
Thermo-mechanical mechanisms	Coffin-Manson $N_f = C_0 \times (\Delta T)^{-n}$ AF (ratio of N_f values per stress cycle, stress/use) $AF = N_{use}/N_{stress} = (\Delta T_{stress}/\Delta T_{use})^n$	• AF = acceleration factor • N_f = number of cycles to failure • C_0 = a materials dependent constant • ΔT = entire temperature cycle-range for the device • n = empirically determined constant • Assumes the stress and use ranges remain in the elastic regime for the materials • The Norris Lanzberg modification to this model takes into account the stress test cycling rate

Table 2.2 (continued)

Mechanism	Model	Assumptions
Creep	$TF = B_0(T_0 - T)^{-n} \exp(E_a/kT)$ AF (ratio of TF values, use/stess) = $((T_0 - T_{accel})/(T_0 - T_{use}))^{-n}$ $\exp([E_a/k][(1/T_{accel} - 1/T_{use})])$	• AF = acceleration factor • TF = time to failure, • B_0 = process dependent constant, • T = temperature in K • T_0 = stress free temperature for metal (~metal deposition for temperature aluminum) • $n = 2 - 3$, (n usually ~5 if creep, thus implies $T < T_m/2$) • E_a = activation energy = 0.5 – 0.6 eV for grain-boundary diffusion, ~ 1 eV for intra-grain • k = Boltzmann's constant – 8.625 10^{-3} eV/K

Table 2.3   Various corrosion models [8]

Model	Form	Terms
Reciprocal exponential model	$TF = C_o \exp[b/RH] f(V) \exp[Ea/kT]$	C_o = arbitrary scale factor, $b = \sim300$, $Ea = 0.3$ eV, $f(V)$ = an unknown function of applied voltage
Power law (Peck) model	$TF = A_o RH^{-N} f(V) \exp[Ea/kT]$	A_o = arbitrary scale factor, $N = \sim2.7$, $Ea = 0.7–0.8$ eV (appropriate for aluminum corrosion with chlorides are present), $f(V)$ = an unknown function of applied voltage
Exponential model	$TF = B_o \exp[(-a) RH] f(V) \exp[Ea/kT]$	B_o = arbitrary scale factor, $a = 0.10–0.15$ per %RH, $Ea = 0.7–0.8$ eV, $f(V)$ = an unknown function of applied voltage
RH^2 (Lawson) model	$TF = C_o RH^2 f(V) \exp[Ea/kT]$	C_o = arbitrary scale factor, (typical value 4.4×10^{-4}), RH = Relative humidity as % (100% = saturated), $Ea = 0.64$ eV, $f(V)$ = an unknown function of applied voltage

Table 2.4   Guideline of use and storage conditions for some major market segments

Major market segment	Indoor	Consumer portable	Other
Operating life	5–10 years	5–10 years	7–25 years
Power on (hrs/week)	60–168	60–168	20–168
Cycles/day	Env. cycle: 1–2 Power cycle: 2–4	Env. cycle: 2–4 Power cycle: 4–6	Env. cycle: 2–4 Power cycle: 2–10
Moisture at low power	30–36°C @ 85–92% RH	30–36°C @ 85–92% RH	30–36°C @ 85–92% RH
Operating temperature (ambient in enclosure)	0–40°C	−18 to 55C	−55 to 125°C
Storage temperature	−40 to 50°C	−40 to 55°C	−40 to 55°C

Lifetime predictions require:

- Knowledge of environmental (operating and non-operating), lifetime of end product, and manufacturing use conditions such as subsequent processing steps (packaging, printed circuit boards).
- End product packaging and application.
- Customer's acceptable failure rate over the lifetime of the product.
- Stress conditions necessary to identify failure mechanisms.
- Acceleration testing and models for lifetime prediction.
- Statistical manipulation of failure distributions in reliability testing.

When using acceleration data to predict lifetimes with acceleration models, one must assume that the shape of the curve is the same in the accelerated condition as in the use condition. The case studies at the end of this chapter illustrate various methods used for reliability lifetime prediction.

2.3.6 Lifetime Units

Failure rates are typically also reported in two popular units, FITS and ppm failure. The unit of FITS is defined as the number of failures in 10^9 device-hours. The ppm unit, which is short for parts-per-million, is always given over a stated time interval. The FITS unit is a rate of failure, while the ppm is a cumulative amount of failures out of a known population over a specific time period. The Chi-Squared method of lifetime (FIT) prediction allows cumulative data collected with specific samples to be applied to the broader population of the same design type, and allows this prediction with zero failures (assuming a constant failure rate $h(t)$ in the bathtub curve in Fig. 2.2). Here χ^2 is the Chi-Squared statistical confidence factor (a constant) that is unique for each confidence interval and number of failures in the testing while SS is sample size. Chi-squared confidence factor charts are typically presented as degrees of freedom versus confidence interval; Table 2.5 contains the statistical confidence factor χ^2 as a function of number of failures (F). To calculate the FIT rate for confidence intervals in addition to 60 and 90%, see [11] where degrees of freedom are (2F+2).

$$\text{FIT} = \frac{\chi^2}{2} \times \frac{10^9}{\text{SS} \times \text{hours}} \tag{2.23}$$

It is important to run accelerated testing properly and bring parts to failure. It is ideal statistically to bring all parts tested to failure, yet this is often an impractical use of company resources. The time to failure of the entire population (or most of it), for one individual failure mechanism, can be modeled for lifetime prediction using the statistical concepts covered in this chapter.

Table 2.5 Chi-squared constants for 60 and 90% confidence intervals

No. of failures	χ^2, 60% Conf.	χ^2, 90% Conf.
0	1.83258	4.60516
1	4.04463	7.779438
2	6.210752	10.644618
3	8.350522	13.361582
4	10.47323	15.987198
5	12.58383	18.54934
6	14.6853	21.064168
7	16.77952	23.541838
8	18.86789	25.989432
9	20.95138	28.411962
10	23.03067	30.81329
11	25.10634	33.19626
12	27.17889	35.563176
13	29.24862	37.915968
14	31.31586	40.256058
15	33.38085	42.584768

2.4 Case Studies

2.4.1 Texas Instruments Digital Mirror Device

Perhaps the field's most successful MEMS reliability story is that of the Digital Mirror Device® developed and manufactured by Texas Instruments for the Digital Light Processing® product. The reliability scientists at Texas Instruments had to start from the very beginning. There was no data on the mirror structures they had designed and fabricated, no acceleration models, and no well understood physics of failure. How does the MEMS reliability engineer start from scratch?

For new MEMS products, the test to failure approach is recommended, coupled with FMEA (Failure Modes and Effects Analysis, see Chapter 5). FMEA is a tool that is used in design and processing of parts, and can be applied to reliability as well. Various methods of collecting data for reliability are highlighted through the FMEA process.

Some background is given on the Texas Instruments product to foster discussion on failure mechanisms and accelerated testing. The failure mechanism we focus on here was hinge memory, also simply known as creep. Creep in metals is a complex mechanism that is a function of stress, temperature, whether the stress is cyclic or steady-state, and the melting temperature of the metal under study. Creep occurs in metals under constant stress and results in plastic deformation.

To determine how the metal will creep, a deformation mechanism map is helpful. The homologous temperature plotted versus the ratio of the shear stress σ_s over the shear modulus μ provides the basis for a deformation mechanism map (Fig. 2.17) [12]. The homologous temperature is defined as the ratio of the test temperature

Fig. 2.17 Deformation
mechanism map of
homologous temperature
versus normalized shear
stress for a Ti-6 wt%/Al alloy
with an average grain size of
100 μm. Reprinted with
permission. Copyright 1991
Springer Science & Business
Media [12]

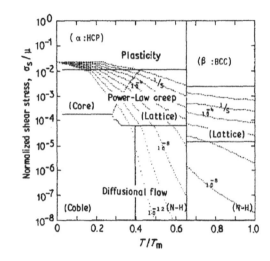

of the material studied to its melting point [13]. If the material under study has
an established deformation mechanism map and it applies to the MEMS structure
geometries, this is a helpful start in understanding and modeling the specific creep
mechanism causing failure. More detail on creep in MEMS is given in Chapter 4,
Section 2.3.

Here, a value of $\sigma_s/\mu < 10^{-4}$ is where diffusional creep occurs. At low
homologous temperatures, Coble creep dominates, and as homologous temperature
increases, Nabarro-Herring creep occurs. These deformation mechanism maps are a
function of the grain size and alloy under study, thus, processing conditions are very
important to creep prediction. Figure 2.17 also has creep strain rates superimposed
over the deformation map. The concept of deformation maps was initially suggested
by Weertman [14–16] and was developed by Ashby and co-workers [17, 18].

Often MEMS products are so unique that standard automated (or manual) test
equipment does not exist (see Chapter 6). Texas Instruments (TI), like many MEMS
developers, had to build their own test stations. Using these test stations, parametric
definitions of population behavior, or goodness of parts, was measured on every lot
(Fig. 2.18).

The TI DMD mirror structure is a hinge/yoke structure (Fig. 2.19). The mirror is
tilted and touches the surface below with a spring tip [19]. This touching action is
called "landing".

The Texas Instruments mirror is made up of various layers. An early depiction of
the structure is outlined next [20].

This is shown for the reader to understand the complexity of the mechanical
structure. The entire assembly sits on a CMOS memory chip. The mirror is not
silicon based as many MEMS structures are, but is aluminum based. The mirror
receives its voltage for electrostatic actuation through the CMOS chip via contacts
etched into the structure. Voltages can be applied that result in the voltage mirror

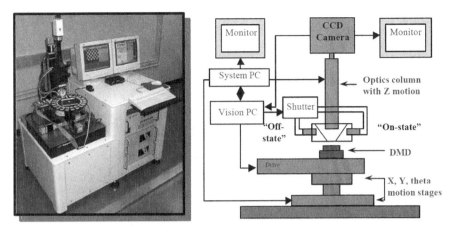

Fig. 2.18 DMD test station. Reprinted with permission. Copyright 2003 SPIE [19]

Fig. 2.19 Illustration of two landed DMD Mirrors. Reprinted with permission. Copyright 2003 SPIE [19]

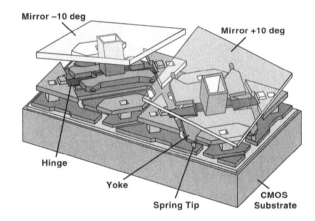

"landing" as seen in Fig. 2.19. Stepped voltages and bipolar reset were included to enhance dynamic control. For more detail on this concept, see references 20 and 21 (Fig. 2.20).

Characterization by bringing parts to failure is a good initial method to determine how a MEMS structure will fail. This method was employed to generate data on which TI mirror designs and processes were most robust. Using the DMD test station, Bias/Adhesion Mirror Mapping (BAMM) Landing Curves were generated [18] upon initial performance and over time (Fig. 2.21). The method of varying one parameter at a time to understand its effect was initially used. Voltage curves were produced by applying voltage to many mirrors and collecting this data to get a distribution of the voltage range for landing. It is likely that voltage was chosen as processing and design variations can result in layer to layer thickness variations, for example, that could result in a range of voltages required for mirror landing. For any unique MEMS design and process, choosing the primary parameters to track is

Fig. 2.20 Exploded view of early TI DMD. Reprinted with permission. Copyright 1998 IEEE [20]

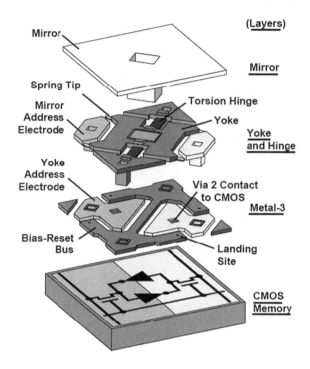

Fig. 2.21 BAMM landing curves example. Reprinted with permission. Copyright 2003 SPIE [19]

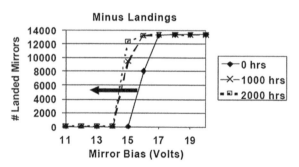

critical to the learning process. Operating parameters are often chosen initially as they need to be specified for field usage.

The change in the landing curve bias voltage over time is a clue to a change in the performance of the device. If the mirror bias changes over time, the voltage applied for landing could move out of the operating range and a failure can occur during the lifetime of the product. Thus, the reliability physics of this mechanism was studied and understood, and acceleration techniques were used to gather data in a faster manner. Figure 2.22 is an example of the work done by the Texas Instrument engineers and scientists in not only understanding the landing curve bias change, but in accelerating it as well.

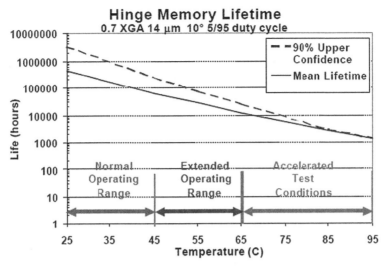

Fig. 2.22 Lifetime curve obtained by Texas Instruments. Reprinted with permission. Copyright 2003 SPIE [19]

Fig. 2.23 Weibull probability plot at worse case duty cycle as a function of temperature, with lifetime predictions at maximum use temperature. Reprinted with permission. Copyright 2002 IEEE [21]

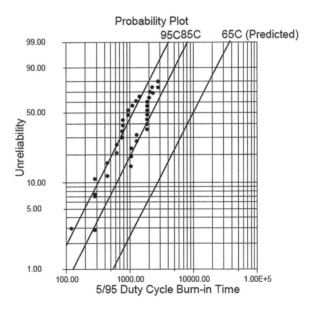

The mechanism for change in the bias voltage was termed "hinge memory" which is a creep mechanism with contribution from surface effects [21]. Creep is known to be accelerated by temperature, thus, stress testing at elevated temperature was performed. Weibull statistics were obtained (Fig. 2.23) and acceleration models were obtained for the failure mechanism. In the case of temperature stresses, the

Arrhenius model is used as the acceleration model and to determine the acceleration factor A_F (Equation (2.24)), this is industry standard in both semiconductors
and MEMS. In using the following model for temperature acceleration, empirical
work must be performed to determine the activation energy for the specific failure
mechanism and materials set. In the case of the hinge memory failure mechanism,
the activation energy (Ea) was ≥ 0.78 eV [21]. In Fig. 2.19, prediction at the maximum operating temperature of 65°C was performed using the model below, and was
compared with data collected through accelerated temperature testing.

$$A_F = e^{Ea/k((1/T_{use})-(1/T_{accel}))} \tag{2.24}$$

The acceleration factor will be calculated. Using equation (2.24), the acceleration
temperatures of 85°C and the use temperature of 65°C, the activation energy of
0.78 eV, and Boltzmann's constant 8.617×10^{-5} eV/°K, equation (2.24) transforms
to (2.25):

$$A_F = e^{0.78 \text{ eV}/8.617E-5 \text{ eV/K}\left(\frac{1}{(273+65)\text{ K}} - \frac{1}{(273+85)\text{ K}}\right)} = 4.46 \tag{2.25}$$

Development of the activation energy is excellent work, yet this acceleration factor is very dependent on the proper activation energy. Table 2.6 shows the change in
acceleration factor when the activation energy is slightly changed. An incorrect activation energy coupled with other use factors (duty cycle is the factor in Fig. 2.23)
could greatly alter operational lifetime predictions.

Table 2.6 Acceleration
factors for various activation
energies using use
temperature of 65°C and
acceleration temperature of
85°C

Activation energy (eV)	Acceleration factor
0.6	3.16
0.65	3.48
0.7	3.83
0.75	4.21
0.78	4.46
0.8	4.64
0.85	5.11

The Texas Instruments DMD example is excellent reliability work performed to
create a niche market for MEMS micromirrors. It also serves as an example to reliability physicists on how to characterize a potential failure mechanism and eliminate
its effects.

2.4.2 Case Study: Analog Devices Accelerometer

The DMD example identifies how to predict lifetime for the DMD design under
worst-case duty cycle as a function of temperature. An example of stiction-based

lifetime prediction as a function of mechanical shock was performed at Analog Devices on test vehicles and will be covered here. The survivor function is used, yet interestingly is a function of shock profile and not a function of time.

The Analog Devices accelerometer family is based on a differential capacitive sensing structure. Fixed and movable beams are adjacent to one another; as a mechanical shock is applied as depicted in Fig. 2.24, the movable beams move which changes the spacings between the fixed and movable beams (air gap capacitor) and results in a unique output voltage. The sensitivity of the MEMS structure, also known as the output voltage per gee-level, is a known value for each accelerometer design. Upon experiencing an externally applied mechanical shock, the shock value and pulse shape can be quantitatively determined with external algorithms (Fig. 2.25).

The Analog Devices MEMS products are amongst the highest reliability MEMS products in the world. With extremely low failure rates in the field, studies like

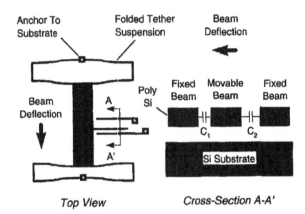

Fig. 2.24 Analog devices differential capacitive sensing MEMS structure. Reprinted from the Lancet. Copyright 1998, with permission from Elsevier [22]

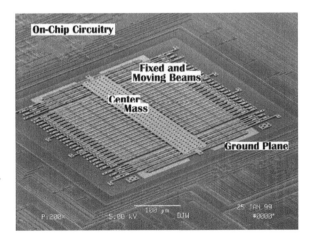

Fig. 2.25 Scanning electron micrograph of the ADXL76 Sensor structure showing the on-chip circuitry, center mass, fixed and moving beams and the ground plane. Reprinted with permission. Copyright 1999 IEEE [23]

Fig. 2.26 Scanning electron micrograph showing z-axis spacing, h. Reprinted with permission. Copyright 1999 IEEE [23]

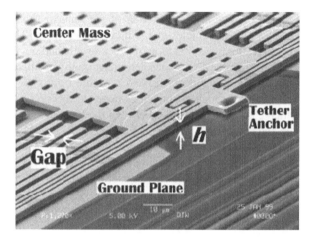

this are very difficult to perform due to the huge sample size requirements. Failures in single digit parts per million levels are typical in the field application for the Analog Devices accelerometer products [24]. In this study, a test vehicle was produced with reduced "h" spacing (Fig. 2.26) to study z-axis stiction. Typically, the restoring force of the structure would exceed the electrostatic and surface forces which would recover the device in the case of structure contact with the ground plane. Yet this test vehicle was more prone to z-axis stiction, and allowed data collection with a realistic sample size. This was an interesting study as a model could be developed to predict failure as a function of a shock profile with a relatively small number of parts.

A mechanical shock test set-up was built to test the test vehicle's stiction behavior. Figures 2.27 and 2.28 are simplified versions of the test set up required for this type of study. A shock was applied to a board-mounted packaged accelerometer test vehicle, a reference accelerometer measured the shock, and the output of the MEMS device was detected with an oscilloscope.

The failure probability of the test vehicle was determined by applying repeated mechanical shocks of various values to the parts. Since the data per part showed that stiction events were not dependent on the shock history of the part, the law of independent probabilities was used to determine the probability of failure as a

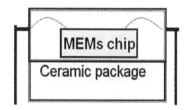

Fig. 2.27 MEMS device in ceramic package

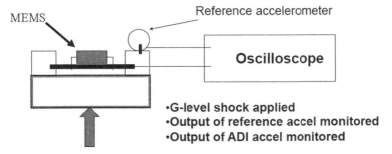

Fig. 2.28 Simplified diagram of test set up for the stiction-susceptible test vehicle

function of a series of shocks. A model was developed to predict the behavior of this population for any shock profile within the experimental mechanical shock range.

The following equation that predicts the number of failures was empirically determined from the test vehicle population [23].

$$F = q_f P_f \{G_z(s)\} \tag{2.26}$$

Here, F is predicted number of accelerometer test vehicle failures as a function of experienced mechanical shocks; q_f is the quantity of stiction-susceptible test vehicles, and $P_f\{G_z(s)\}$ is an empirically determined failure distribution (2.26). $\{G_z(s)\}$ is the z-axis lifetime shock profile that the accelerometer could theoretically experience. The survival function for this study, shown in Fig. 2.30, is related to the probability of failure as described earlier in this chapter. $P_s\{G_z(s)\}$ is the survival function as a function of mechanical shock gee level, versus in Section 2.3, where survival rate is a function of time.

$$P_f\{G_z(s)\} = 1 - P_s\{G_z(s)\} \tag{2.27}$$

$$P_s\{G_z(s)\} = P_{s1}(G_z) \times P_{s2}(G_z) \times P_{s3}(G_z) \times \cdots \times P_{sn}(G_z) \tag{2.28}$$

As an example random series of shocks can be input into the model to obtain an overall failure rate. Equation (2.27) calculates the probability of failure of the shock profile $100g$, $200g$, $300g$, $400g$, and $500g$.

$$P_f\{G_z(s)\} = 1 - [P_s\{100\,g\} \times P_s\{200\,g\} \times P_s\{300\,g\} \times P_s\{400\,g\} \times P_s\{500\,g\}] \tag{2.29}$$

The individual survival rates are taken from the empirical data polynomial fit and a final failure rate is determined.

$$P_f\{G_z(s)\} = 1 - \{(0.9809) \times (0.9636) \times (0.9481) \times (0.9344) \times (0.9225)\} \tag{2.30}$$

$$P_f\{G_z(s)\} = 0.2275 \tag{2.31}$$

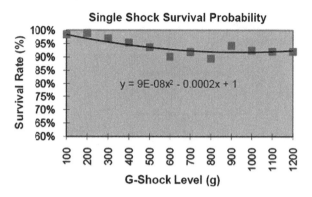

Fig. 2.29 Second order
polynomial fit to mechanical
shock survival data.
Reprinted with permission.
Copyright 1999 IEEE [23]

   In the example shown the probability of failure is 22.75% for a set of test vehicles
that undergoes a random series of shocks. A simple second order polynomial fit
was made to the empirical data to obtain a survival rate for a single mechanical
shock, this was put into the model to obtain the probability of failure for a series
of shocks at various mechanical shock levels (shock profile). Figure 2.29 contains
the mechanical shock survival data. A proprietary acceleration factor would next
be used to apply this model to a product design for stiction prediction. For more
information on mechanical shock physics of failure see Chapter 4 while stiction
is covered in Chapter 3. This example shows how MEMS can change the game;
mechanical failures are not always a function of time. In this case, the survival and
failure rates are a function of operational lifetime mechanical shock profiles.

2.4.3 Case Study: RF MEMS

Compared to existing solid-state technologies for switching 1–40 GHz signals, RF
MEMS switches offer the potential for lower insertion loss, extremely high lin-
earity, and greatly reduced power consumption, in addition to possible integration
with microwave circuits. Since the first RF MEMS switch reported in the 1970s
[25], enormous progress has been made in performance and reliability. Excellent
performance has been demonstrated for electrostatically operated devices [26].
   Yet, unlike MEMS accelerometers and TI's DMD chips, RF MEMS have not
yet found widespread acceptance, and are not a mass produced COTS (component
off the shelf) part. The barrier to commercialization is partially cost, but principally
long-term reliability and packaging. We shall give a brief overview of the operation
principle of capacitive RF MEMS switches, then discuss the main failure mode of
dielectric charging, and the research that has allowed accelerated testing of such
devices.
   This section will focus principally the RF switch first developed at Raytheon
[27], and now being brought to market by MEMtronics Corp., for which some of
the trade-offs necessary for achieving long-term reliability have been published.

There are two mains classes of RF MEMS switches: (a) ohmic switches, in which two conductors are brought into contact to close a circuit, essentially a miniaturized relay switch, and (b) capacitive switches, in which a membrane is moved to change the capacitance between the RF signal line and ground. We discuss only the capacitive switch in this section.

Figure 2.30 is a schematic cross-section of a MEMS capacitive switch. The movable membrane, generally a few hundred nm thick aluminum alloy, can be electrostatically deflected down by a few μm to rest on the thin dielectric covering the central metal conductor. Figure 2.31 provides a top view of a device from MEMtronics, on which the three horizontal conductors (ground-signal-ground) can be seen. The center conductor carries both the RF signal and the DC actuation voltage. When no voltage is applied the membrane remains in the up position, providing a small capacitance C_{off} from the signal line to ground. When a sufficiently high voltage DC is applied to the signal line, the membrane collapses on the dielectric,

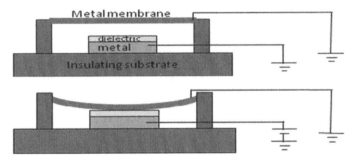

Fig. 2.30 schematic cross-section of a capacitive RF MEMS switch, *top*: undeflected (no dc bias), capacitance C_{off}, *bottom*: snapped down (bias voltage larger than $V_{pull-in}$), larger capacitance C_{on}

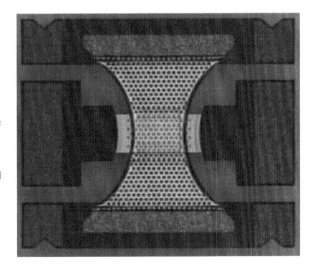

Fig. 2.31 Top view of a MEMtronics Corp RF capacitive MEMS air-gap switch on glass substrate. The membrane is the hour-glass shaped feature. The horizontal central conductor carries both the RF signal and the DC actuation voltage. Reprinted with permission. Copyright 2008 Society of Photo Optical Instrumentation Engineers [28]

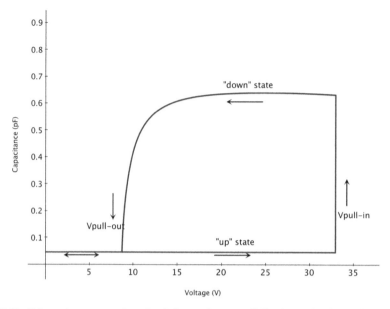

Fig. 2.32 Schematic representation of switch capacitance vs. DC voltage, showing pull-in voltage V_{pi}, pull-out voltage V_{po}, and the hysteresis in operation (which allows using a "hold" voltage much lower than the switching voltage)

providing a much greater capacitance C_{on}. The C_{on}/C_{off} ratio is a good figure of merit. Device operation is illustrated in Fig. 2.32.

The main failure modes reported for this type of switch, associated accelerating factors, and possible solutions are given in Table 2.7.

Stiction will be discussed in Chapter 3, fatigue, creep and dielectric charging in Chapter 4. There is generally a trade-off involved in achieving high reliability, either leading to a small performance drop, or to an increased fabrication complexity, or to a slight increase cost due to a hermetic package for instance. For example, reducing the actuation voltage can provide orders of magnitude increase in device lifetime by reducing charging. However lower voltage operation, assuming we do not change the conductor widths as they need to present the correct impedance, requires a more compliant suspension, which leads to reduced power handling (due to self-actuation), less restoring force and hence higher susceptibility to stiction, and requires a better stress engineering if the metals.

Dielectric charging has been identified as the most important failure mechanism for this type of switch. Switch operation generally require at least 30 V, and for a typical dielectric thickness of 300 nm, this leads to an electric field of 10^8 V/m when the switch is in the down position. At such high fields, charge can readily tunnel into dielectrics and lead to trapped charge. The charge transport mode depends on the dielectric (generally Frenkel-Poole for silicon-rich Silicon nitrides, and Fowler-Nordheim for silicon oxides), but is a complex phenomenon,

Table 2.7 Capacitive RF-MEMS main failure mode and techniques that can accelerate those failure modes

Failure mode	Accelerating conditions	Possible design change
Creep of metal membrane	Temperature, RF power (leading to heating), stress in metal layer	More creep- resistant alloy, better conductor for less ohmic heating
Stiction	Humidity, surface cleanliness, surface roughness	Hermetic packaging, roughness control of membrane and dielectric
Fatigue in membrane	Number of cycles, temperature	Reduce maximum stress by geometry change, change alloy
Dielectric charging	Humidity, electric field, temperature	Lower operating voltage, change dielectric, patterned dielectric, separate signal and drive electrodes

with strong dependence on dielectric composition and deposition conditions, surface cleanliness, geometry.

Charging only occurs in this down position. It was shown that for charging is the total time in the down state, rather than the number of cycles that defines lifetime [29]. Trapped charge leads to failure either from the membrane being stuck down if the trapped charge generates a sufficient electrostatic force, or the membrane being stuck up is the trapped charge screens the applied voltage and raised V_{pi} over the normal operating point.

Dielectric charging in RF MEMS switches takes two main forms: (a) bulk charging due to charge injected into the dielectric from the bottom electrode, and (b) surface charging on top of the dielectric. [28, 30]. Bulk charge leads to a decrease in V_{pi} as the field from the charge adds to the applied field. Surface charge screens the applied voltage, leading to an increase in V_{pi}.

Surface charging is generally avoided at all cost because it shows rapid charging, but slow discharge, and has much higher charge density than bulk charge. Surface charging occurs at voltage above 45 V, and in the presence of humidity or surface contamination. So by operating at lower voltages and in a dry and clean environment, it is possible to limit charging to bulk charging. This is the approach MEMtronics have shown [28].

Using transient current spectroscopy on test structures (metal-insulator-metal structures with no moving parts as well as on working RF-MEMS switch), the MEMtronics team developed a technique to quantify charge tunneling and trapping in the dielectric [26]. This tool allowed the rapid comparison of different dielectrics, as well as the prediction of trapped charge as a function of time, voltage and duty cycle.

Solutions to minimize charging include:

- Increasing the thickness of the dielectric to reduce the applied field, at the cost of lower C_{on}. If one also switches to a dielectric with a higher relative permittivity, as is done for instance for gate dielectric stacks, one could increase thickness without reducing C_{on}. This introduces processing challenges.
- Decreasing the drive voltage. Goldsmith et al. have shown a factor of 10 increase in lifetime for every 5 V reduction in drive voltage [31]. A commonly used technique is a stepped waveform, suing a high voltage for switching and a low voltage for holding the membrane down.
- Change the dielectric to one with less charge traps. For instance MEMtronics uses silicon oxide rather than silicon nitride, which was the "conventional" solution for years for RF MEMS switches. About an order of magnitude reduction in charging is seen with a suitable SiO_x dielectric for comparable C_{on}, as can be seen in Fig. 2.36 [30, 32].
- Pattern the dielectric to minimize the area where charge can accumulate. This is an effective technique, and is implemented in the device in Fig. 2.31. The tradeoff is increased lifetime for decreased C_{on}. [28]
- Hermetic or dry packaging to control humidity. This addresses principally surface charging, and has been successfully implemented [33].

Figure 2.33 shows an example of an accelerated test on a MEMtronics test vehicle, where a 35 V DC signal is used with a 100% duty cycle, showing absolute worst case charging, which appears to be surface charging as V_{pi} increases with time. By way of comparison, Fig. 2.34 is data for a similar device, but under less accelerated conditions, showing no evolution of V_{pi} with time. This type of data allows lifetime to be accurately predicted for well-defined operating conditions.

Controlling surface charge cannot be done without controlling the ambient humidity, which requires a hermetic package. For cost reasons this package must be as compact as possible, and wafer-level packaging is the accepted route, with

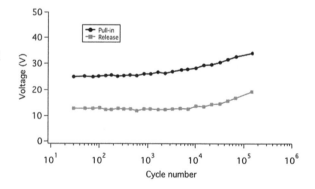

Fig. 2.33 Shift of pull-in and pull-out voltages vs. time for an RF-MEMS switch under accelerated conditions (35 V DC bias, 100% duty cycle). Reprinted with permission. Copyright 2008 Society of Photo Optical Instrumentation Engineers [28]

Fig. 2.34 Shift of pull-in and pull-out voltages vs. time for an RF-MEMS switch under less accelerated conditions than in Fig. 2.33. Reprinted with permission. Copyright 2008 Society of Photo Optical Instrumentation Engineers [28]

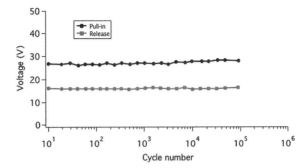

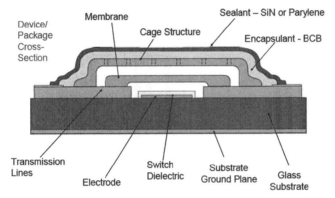

Fig. 2.35 Schematic cross-section of wafer-level packaging developed by MEMtronics to hermetically seal RF MEMS switches with minimal footprint. Reprinted with permission. Copyright 2005 ASME [33]

many different technologies having been demonstrated. The MEMtronics device uses a wafer-level packaging scheme that has a particularly small footprint, shown schematically in Fig. 2.35. The need for such ambient control is illustrated in figure where the effect of %RH is clearly seen as a shift in V_{pi}.

The packaging scheme of Fig. 2.35 was subjected to a number of accelerated lifetime cycles, using elevated temperature and elevated humidity levels to compare different sealants and encapsulants. The tests were highly accelerated (computed acceleration factor of 10^5) and so must be interpreted with care, as higher activation energy mechanisms will be much more highly accelerated, and infant mortality will be overlooked. Nevertheless, the data shown in Fig. 2.37 suggest the package will remain hermetic for over 20 years, thus ensuring surface charging will not occur over the useful life of the RF switch.

To conclude on this example of a capacitive RF MEM switch, very significant progress has been made in identifying failure modes, in determining accurate ways of accelerating those failures, and in devising ingenious solutions to ensure good lifetime while keeping the overall cost down. In-situ diagnostic tools have been

Fig. 2.36 Shift in V_{pi} after
5 min high field stress, for
silicon oxide (*top*) and silicon
nitride (*bottom*) for different
relative humidity conditions,
illustrating both the
superiority of silicon oxide
and the strong effect of
relative humidity on surface
charge. Reprinted with
permission. Copyright 2009
IEEE [30]

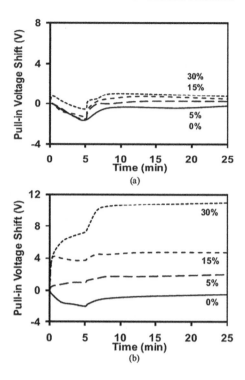

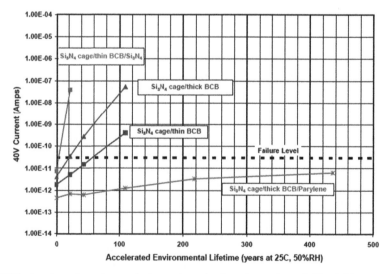

Fig. 2.37 Accelerated test data on the hermetic packaging scheme shown in Fig. 2.36 extrapolated
to normal operating conditions. Reprinted with permission. Copyright 2005 ASME [33]

developed, allowing lifetime to be accurately predicted, and have shown that charging effects can be managed. One can look forward to the commercialization of RF-MEMS switches in the near future.

2.5 Summary

The importance of predictive modeling, acceleration factors, physics of failure, and separating populations for lifetime prediction is covered in this chapter. Examples of failure mechanisms include creep, stiction, and dielectric charging. Accelerated testing and test set-ups are covered through these published examples. Physics of failure is covered in more detail in Chapters 3 and 4, and MEMS test platforms are covered in Chapter 6.

References

1. Guttman, I., Wilks, S.S., Hunter, J.S. (1982) *Introductory Engineering Statistics*. New York: John Wiley and Sons.
2. Nash, F.R. (1993) *Estimating Device Reliability: Assessment of Credibility*. Dordrecht: Kluwer Academic Publishers, p. 64, Springer Publishing, now copyright holder.
3. Klinger, D., Nakada, Y., Menendez, M. eds. (1990) *AT&T Reliability Manual*. New York: Van Nostrand Reinhold.
4. Dodson, B. (2006) *The Weibull Analysis Handbook*, 2nd edn. Milwaukee: American Society for Quality, Quality Press.
5. Tanner, D.M. et al. (2000) *MEMS Reliability: Infrastructure, Test Structures, Experiments, and Failure Modes*. Sandia Report SAND2000-0091, p. 78 (Courtesy Sandia National Laboratories, Radiation and Reliability Physics Dept., www.mems.sandia.gov)
6. Courtesy of Sandia National Laboratories. SUMMiT(TM) Technologies, www.mems.sandia.gov, http://mems.sandia.gov/gallery/images.html
7. Shea, H.R. (2006) Reliability of MEMS for space applications. In V. D.M. Tanner, R. Ramesham (eds) *Reliability, Packaging, Testing, and Characterization of MEMS/MOEMS*. Proc. of SPIE Vol. 6111, 61110A.
8. Blish, R., Durrant, N. (2000) Semiconductor device reliability failure models. Int. Sematech. Technol. Transfer # 00053955A-XFR, May 31.
9. Blish, R., Huber. S., Durrant, N. (1999) Use condition based reliability evaluation of new semiconductor technologies. *International Sematech* Technology Transfer # 99083810A-XFR, August 31.
10. Planning, Developing and Managing an Effective Reliability and Maintainability (R&M) Program (1998) NASA-STD-8729.1, National Aeronautics and Space Administration, December 1998.
11. FLA (1986) *Standard Mathematical Tables*, 26th edn. Boca Raton: CRC Press, p. 548.
12. Janghorban, S. et al. (1991) Deformation-mechanism map for Ti-6 wt% Al Alloy. J. Mater. Sci. 26, 3362–3365.
13. Dieter, G.E. (1986) *Mechanical Metallurgy*. New York: McGraw-Hill, Inc.
14. Weertman, J. (1956) J. Mech Phys Sol. 4, 230.
15. Weertman, J. (1960) Trans. AIME 218, 207.
16. Weertman, J. (1963) Trans. AIME 227, 1475.
17. Ashby, M.F. (1972) Acta Met. 20, 887–897.
18. Frost, H.J., Ashby, M.F. (1982) Deformation-mechanism maps. New York: Pergamon Press.

19. Douglass, M.R. (2003) DMD reliability: a MEMS success story. In R. Ramesham, D. Tanner (eds) *Reliability, Testing and Characterization of MEMS/MOEMS II*. Proceedings of SPIE Vol. 4980, SPIE.
20. Douglass, M.R. (1998) Lifetime estimates and unique failure mechanisms of the digital micromirror device (DMD). Reliability Physics Symposium; 1998 IEEE International Volume, Issue 31.
21. Sonheimer, A. (2002) Digital micromirror device (DMD) hinge memory lifetime reliability modeling. IEEE 40th Annual International Reliability Physics Symposium, Dallas Texas.
22. Chau, K., Sulouff, R. (1998) Technology for the high-volume manufacturing of integrated surface-micromachined accelerometer products. Microelectron. J. 29, 579–586.
23. Hartzell, A., Woodilla. D. (1999) Reliability methodology for prediction of micromachined accelerometer stiction. 37th International Reliability Physics Symposium (IRPS), San Diego, California. p. 202.
24. Hartzell, A. et al. (2001) MEMS reliability, characterization, and test. In R. Ramesham (ed) *Reliability, Testing, and Characterization of MEMS/MOEMS*. Proc. SPIE Vol. 4558, pp. 1–5.
25. Peterson, K. E. (1979) Micromechanical membrane switches on silicon. IBM J. Res. Develop. 23(4), 376–385, July 1979.
26. Yuan, X., Hwang, J.C.M., Forehand, D., Goldsmith, C.L. (2005) Modeling and characterization of dielectric-charging effects in RF MEMS capacitive switches. IEEE Int. Microwave Symp. paper WE3B-3, June 2005.
27. Yao, Z.J., Chen, S., Eshelman, S., Denniston, D., Goldsmith, C. (1999) Micromachined low-loss microwave switches. J. Microelectromech. Syst. 8(2), 129–134, June 1999.
28. Goldsmith, C.L., Forehand, D., Scarbrough, D., Peng, Z., Palego, C., Hwang, J.C.M., Clevenger, J. (2008) Understanding and improving longevity in RF MEMS capacitive switches. Proc. Int. Soc. Optical Eng. 6884(03), Feb 2008
29. Van Spengen, W.M., Puers, R., Mertens, R., De Wolf, I. (2004) A comprehensive model to predict the charging and reliability of capacitive RF MEMS switches. J. Micromech. Microeng. 14(4), 514–521.
30. Peng, Z., Palego, C., Hwang, J.C.M., Moody, C., Malczewski, A., Pillans, B., Forehand, D., Goldsmith, C. (2009) Effect of packaging on dielectric charging in RF MEMS capacitive switches. IEEE Int. Microwave Symp. Dig., 1637–1640, June 2009.
31. Goldsmith, C., Ehmke, J., Malczewski, A., Pillans, B., Eshelman, S., Yao, Z., Brank, J., Eberly, M. (2001) Lifetime characterization of capacitive RF MEMS switches. IEEE Int. Microwave Symp. 1, 227–230, May 2001.
32. Peng, Z., Palego, C., Hwang, J.C.M., Forehand, D., Goldsmith, C., Moody, C., Malczewski, A., Pillans, B., Daigler, R., Papapolymerou, J. (2009) Impact of humidity on dielectric charging in RF MEMS capacitive switches. IEEE Microwave Wireless Comp. Lett. 19(5), 299–301, May 2009.
33. Forehand, D.I., Goldsmith, C.L. (2005) Wafer level micropackaging for RF MEMS switches. ASME InterPACK '05 Tech Conf, San Francisco, CA, July 2005.

Chapter 3
Failure Modes and Mechanisms: Failure Modes and Mechanisms in MEMS

3.1 Introduction

As defined in Chapter 2, reliability engineering is the process of analyzing the expected or actual failure modes of a product and identifying actions to reduce or mitigate their effect. A *Failure Mode* describes the way in which a product or process could potentially fail to perform its desired function and can be defined in several ways, of which the most common is in a progression of time, where a failure mode comes between a *cause* and an *effect*. However, it is also possible that in some cases the cause or effect themselves might be the failure mode or for a single event to be a cause, effect and a failure mode. In practice, it is more likely that a single cause might have multiple effects or a combination of causes might lead to an effect. Failure modes are sometimes also called *categories of failures* and may be broadly categorized into two types – *Design Failure Modes and Manufacturing Failure Modes*, depending on their origin in the product development phase.

In this chapter, we take a broad look at defect mechanisms and associated failure modes typically observed in a variety of MEMS enabled products, without necessarily limiting ourselves to a specific type of fabricated device. The chapter focuses on failure modes observed primarily in the MEMS element, and to a lesser extent failure modes related to the interaction of the sensor and the package, but not specifically to the IC that may co-exist with the sensor in the product. A more detailed study of IC related failure is wide available [1] and is usually very specific to the process technology node. One key characteristic that differentiates MEMS sensors from traditional IC's is the use of a wide variety of unique process steps and engineering materials. The focus will not be on specific micromachining process steps used to create the MEMS element (such as in [2]) but on failure modes encountered in such processing and which are dependent on the particular product being developed. Surface micromachining for example, is one of the most popular fabrication flows for MEMS which uses a doped silicon starter wafer with subsequent layers of polysilicon, oxide, nitride and metal such as shown in Fig. 3.1. Such MEMS are fabricated with a wide variety of materials including metals, dielectrics [3], and polymers which are not as unique as silicon [4] in terms of being used for both the electrical and mechanical parts of a MEMS element [5], and has been

A.L. Hartzell et al., *MEMS Reliability*, MEMS Reference Shelf,
DOI 10.1007/978-1-4419-6018-4_3, © Springer Science+Business Media, LLC 2011

Fig. 3.1 MEMS assembly
fabricated in polysilicon
(reprinted with permission
Copyright 2005 Simon
Frasier University – Institute
of Micromachine and
Microfabrication Research
[6])

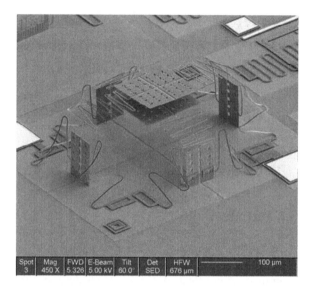

shown to be versatile not only for designers but because it is highly conformable
to standard manufacturing processes used in the semiconductor industry and thus
reduces the need to develop completely new fabrication infrastructure.

As mentioned earlier, most failures observed in any MEMS enabled product can
usually be traced back to a design or manufacturing decision [7], and the choice
made to either use a particular material or process step. In a given product, the
design phase may introduce failure mechanisms of three different varieties – *functional, material,* or *non-analyzed* depending on whether the design was properly and
sufficiently analyzed for the chosen fabrication process and operating conditions or
not, and whether the particular process chosen has been properly characterized. This
has not always been easy because the earliest MEMS products relied heavily on a
methodology of successive iteration[1] to achieve the performance functionality necessary, but in recent years, there has been a significant increase in the use of MEMS
design tools that can provide detailed insight into the behavior MEMS devices prior
to actual fabrication. While this has significantly reduced the overall product development time and cost, there has been a significant impact in the reduction of the
possible failure modes or mechanisms.

The next section discusses potential failures modes that originate in the design
phase.

---

[1] See description in Chapter 1 – see Figure 1.4

3.2 Design Phase Failure Modes

The discussion in this section is restricted to failure modes that are distinguished by their origin in the design phase where they mainly impact the reliability of the product through the performance of the device. Broadly, we can identify two sub-categories of failures – *functional and material modes*.

3.2.1 Functional Failure Modes

Functional failure modes imply a degradation, loss or absence of intended performance under operating conditions due to inadequate or insufficient design leading to a deviation from the product specification. A functional failure affects functionality of the device in the field and thus impacts the overall reliability of the part. The loss of function of the part may occur at the beginning or later in life[2] but either way the failure is due to insufficient design. A good example of such a functional failure mode is the catastrophic failure of the device due to mechanical shock.

In Fig. 3.2, the catastrophic failure of a polysilicon spring is shown. Such a failure mode can routinely occur in MEMS devices either due to inappropriate handling of the part during assembly or if the device is subject to a large enough in-use shock in the field. The behavior of silicon during high-g (shock)[3] events depends on a variety of factors but fundamentally it comes down to the strength of the material and the level of the shock the structure was designed or analyzed for including appropriate assumptions of corner cases. In order to accurately predict failure, a failure criteria is necessary, and the choice of either quasistatic or rate-dependent criteria becomes important. In one study [9], the quasistatic fracture strength was cited as being a valid criterion for the dynamic performance of a MEMS device. However, for more accurate modeling of such failure processes it may be necessary to also

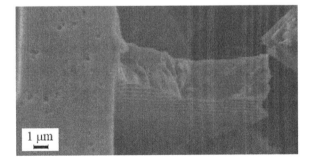

Fig. 3.2 Shock induced failure of a polysilicon spring – Reprinted with permission Copyright 2009 – Sensors MDPI [8]

1 µm

---

[2] see Bath Tub curve Figs. 2.2 and 2.3 – Section 2.3.1.

[3] Refer to Section 4.2.2 for a more detailed discussion on shock.

consider the effects of grain morphology [8], surface roughness, and defect distribution. Additionally, in electro-mechanical devices, there is the further complication of a pull-in instability [10], which has been observed (under these dynamic conditions) to complicate matters sufficiently, producing complicated failure modes [11].

The capability to predict these types of failures is essential to minimizing the risk of field failures and improving the reliability of the part.

The functional failure modes are divided between MEMS element design, system level design and package design. It is possible to also have functional failures due to design of the conditioning circuitry but this is not covered in this book.

3.2.1.1 Element Design

Elemental Design failures include mask data faults, design rule violations, and engineering analysis faults that lead to failures where the MEMS element does not perform as expected.

Mask Data Faults

Mask data faults are fairly common in MEMS design because of the nature of MEMS fabrication process flows which are usually exclusively developed for a specific device. This makes it difficult to create comprehensive design rule checkers (DRCs) that are capable of catching each and every flaw in the mask set used with a particular process flow, and it is not uncommon to have manually executed layout reviews that are time consuming and prone to errors.

Another, unique issue with MEMS design is the use of different CAD layout tools and formats. Historically, it is fairly common to employ multiple formats (DXF, GDS etc.) for handling CAD data and data translation from one format to another can also introduce faults. Additionally, another potential source of flaws in a MEMS mask design is the fact that MEMS devices often will use a non-Manhattan shape such as a circular mass or a curved spring, and semiconductor CAD tools are not completely equipped to handle such shapes. As a result of all of these issues; it is common for faults in the mask design to occur. A good example of such a flaw is shown in Fig. 3.3.

The effect of such mask data faults can be quite serious from a reliability standpoint because in some case this may result in an incomplete etch or over etch which introduces a structural flaw in the MEMS element. Such a flaw could be initially benign but can manifest itself in the field [12].

CAD Models

In a MEMS process, materials are deposited and etched onto non-ideal geometries with complex inter-layers due to particular process sequences. The ability to accurately simulate and predict the functionality of a part in 3D depends largely on the accuracy of the CAD model representation. Most solid modeling tools use a

Fig. 3.3 Typical Mask layout faults (**a**) design rule violations may occur when the shape of the MEMS structure changes, and (**b**) misplacement of parts of the layout

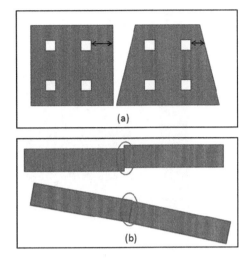

Fig. 3.4 CAD solid model representation of a MEMS accelerometer (reprinted with permission Copyright 1997 Analog Devices)

3D representation called *nurbs*[4] to realize specific shapes but since these are idealized representations the resulting models are characterized by flat surfaces and sharp edges as shown in Fig. 3.4. For surface micromachined structures, such as those depicted in the Fig. 3.4, this is not that inaccurate and with a structured design methodology that examines behavior at the process and property corners, it is possible to bound the behavioral performance of the sensor.

However, MEMS designers create ever complex designs where it is more of a challenge to capture precise geometrical features, making it more likely that the

[4] NURBS: *Non-Uniform Rational B-Splines*, are mathematical representations of 3-D geometry that can accurately describe any shape from a simple 2-D line, circle, arc, or curve to the most complex 3-D organic free-form surface or solid.

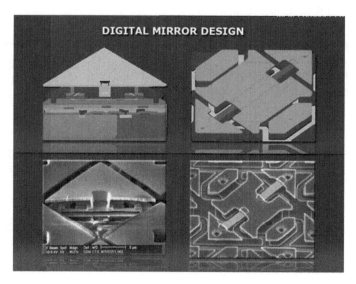

Fig. 3.5 Comparison of a 3D rendered model and an actual MEMS device (reprinted with permission Copyright 2008 – Coventor Inc.)

predictive capability of numerical simulations is limited by the accuracy of the CAD solid model. From observation of the final device we know that a realistic representation of the device would not be possible using the same geometrical representation described, and so more recently, voxel based tools [13] have begun to tackle this complexity and produce more realistic CAD models. In the Fig. 3.5, one can see that these models capture much more details of the real device.

Material Properties

In the design phase, material properties are essential quantities to properly analyze the behavior of the device and the relative inaccuracy of these properties often leads to another type of functional failure – due to inaccurate material properties. Even though the modulus and density of most material used in MEMS are widely available [14], process dependent properties such as residual stress, stress gradient, fracture strength, fatigue limit, and others are not simple to measure and it is fairly common practice for the designer to simply use bulk properties[5] during the design phase. The *residual stress or stress gradient* within a thin film originates from either *intrinsic* or *extrinsic* sources. Intrinsic sources include material phase change, grain growth, crystal misfit, and doping whereas extrinsic sources include plastic deformation, thermal expansion and external loads. The material properties of a thin film can be quite different from bulk properties.

---

[5] Section 7.3.1.2 for summary of material property references and metrology.

Fig. 3.6 Curvature in a
released sensor is a result of
thin film stress which is
highly process dependent –
Reprinted with permission
Copyright 1997 – Analog
Devices

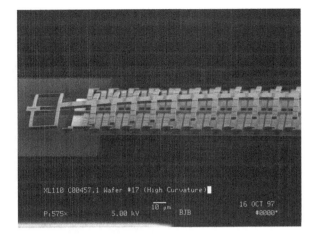

The lack of accurate thin film properties at the beginning of the design effort
can lead to several functional failure modes, where the device does not perform as
expected. In Fig. 3.6, the performance of the comb drive is highly dependent on the
initial curvature within the released polysilicon layer. The stress gradient produces
curvature in MEMS accelerometers that can result in a degraded offset performance
or the part is no longer within specification, or worse still, stress relaxation may
cause the part to gradually drift out of specification over life. Finally, the lack of
accurate reliability related material property data makes it challenging to predict
field reliability of a device [15].

Analysis and Simulation

Another type of failure that may occur is due to insufficient design analysis of the
particular MEMS element. MEMS design tools today are highly specialized analysis
tools that are capable of directly accepting a mask layout file, converting it into a 3-D
numerical simulation model based on the process flow and incorporating all relevant
and necessary material properties [16]. A variety of full-field simulators such as
finite element analysis (FEA) or boundary element analysis (BEA) can then simulate
the device behavior under prescribed boundary and initial conditions encountered
during operation of the device. In the case of a common element such as a comb
drive used in an accelerometer, gyro or resonator, one analysis of interest might
be the deflection behavior due to process induced stress gradients[6] as shown in
Fig. 3.7.

Simulation tools are routinely used to analyze very complex conditions encountered in MEMS devices, such as fluid structure interactions in ink-jets, fluid-
chemical analysis in bio-MEMS, etc. However, the most commonly encountered

---

[6] Additional discussion in Section 5.5.10.

Fig. 3.7 Simulated
displacement due to applied
stress gradient in a capacitive
accelerometer (reprinted with
permission Copyright 2000 –
Coventor Inc.)

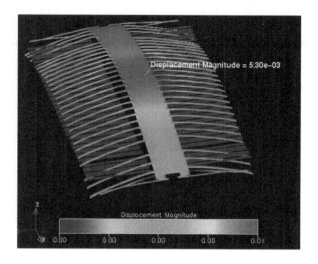

Fig. 3.8 Comparison of
modeled pull-in behavior of
two similar FEA models with
an analytical model (reprinted
with permission Copyright –
Coventor Inc.)

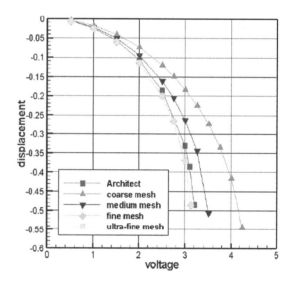

analysis in MEMS is a simple electrostatic *pull-in analysis* for electro-mechanical
devices. If proper care is not taken, it is possible for these predictions, to system-
atically under or over-predict the pull-in voltage that is important to the overall
function of the product. As one can see in Fig. 3.8, the pull-in voltage modeled by
the common coupled finite element-boundary element approach can be over predicted by as much as 20% (or more) if the model is not sufficiently populated with
enough elements [17]. Although some failures may be caught at the fabrication of
the first prototypes, there are other similar simulation analyses that are difficult to
predict until they manifest themselves in the field. One such example of a failure

is the pull-in behavior of an RF switch where dielectric charging causes the pull-in voltage to vary over time [18]. Models for charge accumulation within dielectrics may not be included during the design phase.

Design analysis that are not accurately predicted either for lack of time; simulation tool capabilities or physical understanding can lead systematic non-performance or drift of a single parameter during field operation resulting in a failure.

3.2.1.2 System Level Design

Another design limitation that is routinely encountered which may lead to field failures is related to *system complexity*. A MEMS product is a complex system comprised of MEMS element, electronics, and package, and it is a significant modeling challenge to be able to predict overall system behavior without simplifying the sub-components to a sufficient level of abstraction without loss of accuracy. The system level models may then not give the designer enough predictive information to identify a potentially serious failure mode. Essentially failure predictability decreases as system level model abstraction increases. The Texas Instruments DLP© product, which contains over a million individually addressable mirrors (1024×1024 pixels) with signal processing at each pixel and a custom package[7] is a good example of system complexity. One way to understand the problem of "sufficient analysis" is the fact that for each DMD chip the failure rate is defined or set to be <1 ppm or less than 1 mirror per chip. This requires that the modeling used to predict overall functional performance of the chip have to be of extremely high fidelity [19].

Design Integration

Usually, a MEMS product comprises of a MEMS sensor and control circuitry described in the block diagram in Fig. 3.9. The ability to co-simulate the behavior of the entire system including the sensor can be another significant challenge considering that circuit level simulators are usually not capable of adequately representing the MEMS element [16].

Typical failure modes that could arise due to a lack of such integrated design capability fall into two main categories:

1. Process corners: During the circuit design phase, designers simulate the performance of the ASIC at all process corners and temperatures. During this stage of the design process, the inability to adequately represent the MEMS element over the element's own process corners or behavior over temperature, within the design can lead to multiple failures that affect both functionality and reliability.
2. Circuit Timing: Again during the circuit design phase, it is critical to get the circuit timing for drive and sense signal chains precisely synchronized. If a suitably

---

[7] See Section 2.6.1 for a case study on mirror operating life.

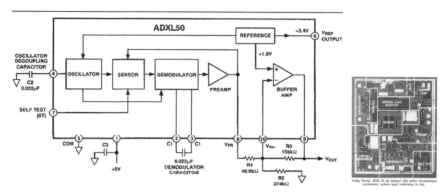

Fig. 3.9 Control system and circuitry for XL50 – (reprinted with permission Copyright – Analog Device)

sophisticated sensor model cannot be integrated into the system level model, this may result in timing failures which are sometimes extremely difficult to pin-point and lead to field failures under certain use conditions.

In recent years, substantial advances in CAD tools have enabled designers to co-simulate both the MEMS and the circuitry required to drive it, in a single environment [16] leading to more robust simulation of the system as a whole.

3.2.1.3 Package Design

Assembly or packaging of MEMS devices present a unique challenge for MEMS designers and reliability engineers alike (Fig. 3.10). The similarities with conventional IC packaging is clear in the sense that the package must reject certain inputs like moisture, contamination, etc. and tolerate common forces such as temperature, shock, handling or tester forces. However, unlike conventional IC packaging, the

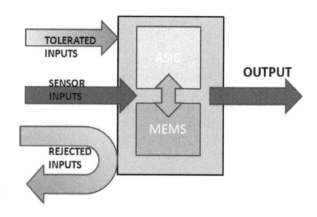

Fig. 3.10 MEMS packaging challenge

MEMS element must interact directly with the outside world in order to perform their design function. In the case of accelerometers, this does not require the package to be designed markedly different than a conventional IC package but in the case of wide variety of other MEMS products which sense pressure, temperature, chemical species, or control light or sound, there needs to be direct interaction between the sensing element and the input that needs to be measured.

Each MEMS application usually requires a new package design to optimize its performance or to meet the needs of the system. This is the primary reason why the cost fraction of packaging a MEMS device remains high [10]. There are several categories of MEMS packages including metal packages, ceramic, plastic packages, and thin-film multilayer packages [20] that is similar to standard semiconductor packaging. However, there are several factors that make MEMS packaging more complicated and establish the need for comprehensive MEMS-package design integration [21]. Some of these factors are:

- Usually might include a discrete circuit chip besides the MEMS device.
- Contains a hermetic bonded silicon cap over the sensor structure which is sensitive to assembly forces.
- Assembly material selection depends on the application and is complicated because of the differences in properties such CTE, modulus, and glass transition.
- Package failure modes observed and reliability issues can be quite diverse [22].

As a result of such factors there are two basic categories of failure modes related to the assembly process – related to *package materials*, and *sensor-package interaction*.

Package Materials

There are a variety of packages used in the packaging of MEMS sensors [15] and these contain an even wider variety of materials include metals, plastics, ceramics, and polymers of various kinds, that all have to function together to ensure performance of the life of the sensor. As shown in Fig. 3.11, a typical MEMS accelerometer may be packaged in a plastic over-molded package surrounded by

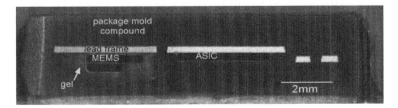

Fig. 3.11 Cross-section of a typical plastic overmolded MEMS sensor – (reprinted with permission Copyright 2009 Chipworks [24])

plastic mold compound, gel, and die attach all of which have different material properties [23]. The reliability of the product depends as much on the package materials as it does on the MEMS element, and several potential failure modes have been observed due to package material selection and characteristics.

These failures may be broadly divided into two types:

1. Interfacial: Interfacial failures arise primarily due to differences in strength or CTE between adjacent materials or poor interfacial adhesion between the layers. In many packages, it is possible to find interfaces such as the one showed in Fig. 3.12, which clearly shows that the die-attach (DA) thickness varies and may even have poor adhesion in certain areas.

Fig. 3.12 Quality of die attach bond line for a MEMS die attached to a substrate (reprinted with permission Copyright 2008 – Analog Devices)

The failure mechanisms induced due to poor interfaces are almost always related to the stress on the MEMS element. Depending on the package and the applied stress state the sensor element can react quite differently but it is clear that in such cases the long term performance of the sensor is affected.

2. Bulk: Material failure modes that arise due to the behavior of package related materials again are almost always stress inducing on the MEMS element. One important property is the glass transition[8] temperature (T_g) of the plastic used in overmolded plastic packages. In a MEMS product such as that shown (in Fig. 3.11) there is a clear interdependence of the performance of the plastic and the reliability of the product. The T_g of some typical polymers used is in

---

[8] Glass Transition Temperature (T_g) – describes the temperature at which amorphous polymers undergo a second-order phase transition from a rubbery, viscous amorphous solid, or from a crystalline solid to a brittle, glassy amorphous solid. for a case study on mirror operating life.

the 120–150°C range but may depend on the cure time i.e. a longer cure time increases the cross-linking and results in a higher T_g. The higher the T_g the better the performance of the part over the operational temperature range which is typically 85°C. For example, in [25] the gyro package shows large nonlinear behavior over the temperature range up to 140°C due to the fact that the mold compound goes through a glass transition. However increasing the T_g might also increase the susceptibility to package cracking during reflow [26].

Other typical defects in the die attach (see Fig. 3.13) include voids, interfacial voids and delamination, and cracks, and for plastic over-molded packages additional defects include wire sweeping, incomplete filling, cracking, blistering and flashing. All of these defects can result in a variety of reliability related failure modes including drift, and component failure. The integrity and strength of the package materials used play a crucial role in the overall reliability of the part.

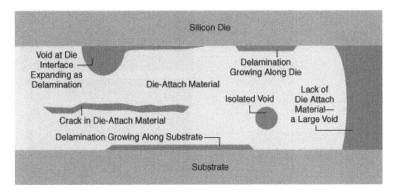

Fig. 3.13   Die attach failure modes

MEMS – Package Interaction

MEMS by their nature require application specific packaging and since the package immediately surrounds the MEMS sensor it has a direct effect on its thermal-mechanical behavior, environmental compatibility and contamination. A major contributor to increased product development cycles is the lack of focus on the package early on in the design phase. It is therefore critical that during the design phase a thorough study of the influence of the packaging on performance be conducted and that this occurs simultaneously with the sensor element design. There are several valid methodologies [27, 28] that depend on the relative size of the package, sensor size and the level of detail required in the particular analysis. This size difference has been known to create a serious challenge for numerical simulators attempting to perform brute force simulations of the MEMS and package together.

The coupling between the package and sensor chip is most commonly observed with temperature effects where the difference in CTE between the mold compounds, die-attach and sensor chip leads to a complex stress state at the sensor. In a MEMS gyro packaged in a plastic overmolded package [25] the local CTE mismatch produces a convex bending of the package at the maximum of the temperature range, but this is of opposite curvature at temperatures lower than 125°C and at room temperature. The MEMS gyroscope in this case must be designed to be less sensitive to these strains as they deform the spring elements of gyroscope, leading to resonant frequency changes of the sensing and driving modes. In order to increase robustness of the gyro to this type of deformation, the designers in this case modified the spring design to reduce the frequency shift observed across the operational temperature range (Fig. 3.14).

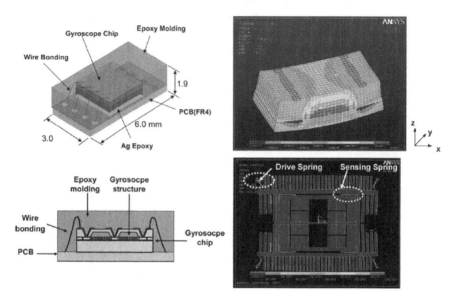

Fig. 3.14 MEMS Gyro packaged in a plastic over-molded package – Reproduced with permission Copyright – 2007 IEEE [25]

The MEMS designer needs to be able to account for similar effects due to hygroscopic swelling of the mold compound [29] and the die attach [30] both of which have long term affects on the behavior and reliability of the part.

In the next section we will take a closer look at some common material failure modes within the sensor element itself. However, the user is referred to Chapter 4 for more details on specific failure modes that manifest in the field.

3.2.2 MEMS Material Failure Modes

Specific material failure modes in MEMS devices can be varied and highly process dependent but are commonly divided into the following categories: *Thermomechanical failures, Electrical Failures, and Environmental.*

3.2.2.1 Thermo-Mechanical (TM) Failures

Thermo-mechanical failures are those failures resulting from thermo-mechanical forces and generally include the most common of MEMS stress failures i.e. residual stress:

1. *Contact Wear*: RF switches or contact actuators [31] are Class III-IV[9] MEMS devices where contact of proximal surfaces (as shown in Fig. 3.15) occurs frequently and with sufficient force to cause time dependent damage resulting in wear of the contact surfaces [32]. During normal operation of the device, these surfaces come into repeated contact and the material in the contact zone is subjected to large stresses under conditions of large current densities and temperature, which eventually lead to *wear* failures. The reliability of the switch is dependent on the material properties and processing conditions of the contact zone [33].

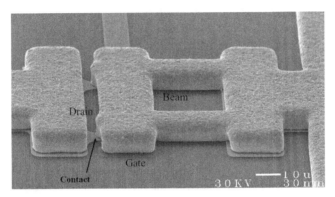

Fig. 3.15 RF MEMS switch contact (reprinted with permission Copyright – NorthEastern University)

Other examples of RF switch reliability maybe found in an interesting case study in Chapter 2, and the wear of the SAM coating on aluminum surfaces in the DMD® mirror maybe found in Section 5.5.7 (AFM Methods). The wear observed in the latter case is purely under mechanical contact conditions that are less extreme than those encountered in an RF switch.

In the case of RF switches, the contacts needs to be able to transmit a current of sufficient magnitude in a very small area resulting in very large gradients of temperature and stress which cause local damage. As a result of this accumulated damage to these interacting surfaces, the contact resistance gradually increases over life until eventually the contact breaks down, as shown in Fig. 3.16.

---

[9] See Fig. 6.2.

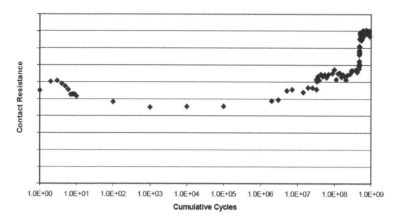

Fig. 3.16 Lifetime data for a typical RF switch (reprinted with permission Copyright – 2002 MANCEF [34])

The requirements to improve wear resistance (and consequently the reliability of the device), are low adhesion, high current capacity and low contact resistance [31]. Contact resistance (assuming an elastic-plastic contact [35]) is made up of two main terms, the constriction resistance (R_c) and the film (tunneling) resistance (R_t) which depend on the hardness (H), force (F), resistivity (ρ), film resistivity (ρ_t), and elastic plastic factor (ζ) described in Equation (3.1).

$$R = R_c + R_t = 0.89\rho \left(\frac{\xi H}{nF} \right)^{1/2} + \frac{\rho_t \xi H}{F} \tag{3.1}$$

To create a lower stable contact resistance it becomes necessary to increase the contact force (F) but this also increases the adhesion force between the surfaces which then requires more force to break the contact, and so is not really an optimal approach [35]. The relationship between the adhesion force and contact force is determined by the nature of the contact i.e. elastic or elastic-plastic, as well as the occurrence of contact heating and welding. The adhesion factor which is the ratio of the separation (breaking) force to the contact force is given by:

$$\frac{F_{\text{separation}}}{F_{\text{contact}}} = \frac{\zeta_1 \zeta_2 \zeta_3}{2} \tag{3.2}$$

where ζ_1 is the slide factor (1–1.5), ζ_2 is the elastic de-compression factor (0.6–1.0), and ζ_3 is the film factor (0–1.0). For pure (bulk) gold (99%), this factor is 0.68 while for ruthenium (*Ru*) coated contacts this is ~0.22. A lower adhesion factor means a lower breaking force which will produce less damage at the contact surfaces. Figure 3.17 shows the contact resistance as a function of contact force for several metallic materials like gold (*Au*), and rhodium (*Rh*).

The force to break the contact is determined by the adhesion and welding forces and it is this breaking force which eventually causes wear and degradation of the

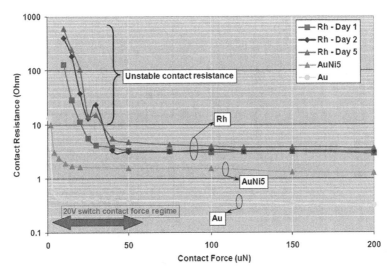

Fig. 3.17 Contact resistance as a function of contact force for several metals (reprinted with permission Copyright – 2007 SPIE [36])

contacting surfaces. For example, the difference in the breaking force between evaporated gold and sputtered gold is an order of magnitude leading to longer lasting contacts from evaporated gold [35].

Another factor in such types of asperity contacts is the temperature rise in the contact which is predicted by the following Equation (3.3):

$$\frac{I^2}{F} = \frac{16L\left(T^2 - T_A^2\right)}{\pi\, H \rho_A^2 \left[1 + \frac{2}{3}\alpha\,(T - T_A)\right]^2} \frac{n}{\xi} \tag{3.3}$$

where L is the Lorenz constant, n characterizes the surface condition, ρ is the resistivity, H is the hardness and T_A and T are the ambient and melting temperatures respectively. The decrease in life cycles between hot and cold switching is observed in Fig. 3.18.

2. *Fatigue*: Fatigue is the collective name for multiple phenomena that arise due to different mechanisms in brittle and ductile materials resulting in a progressive decline in load bearing capacity eventually leading to catastrophic failure. These types of failures are particularly troublesome because of the process dependent behavior of thin film materials over many accumulated cycles of stress. Cyclic fatigue damage may cause several types of performance failures such as resonant frequency decreases (Section 5.5.8), drift and catastrophic failure. The key design parameter for fatigue is the Endurance Limit (S_m) but some MEMS materials such as aluminum do not have a well defined endurance limit, while others

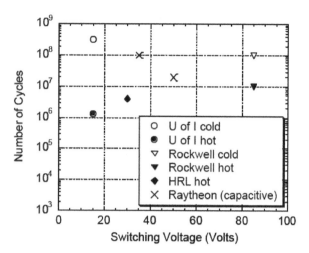

Fig. 3.18 Switch lifetime vs. actuation voltage several published works [37–39] (reprinted with permission Copyright – 2002 MANTECH [33])

such as silicon or *SiN* are not known to exhibit fatigue at typical operational levels. A more detailed discussion of fatigue mechanisms in MEMS materials is available in Section 4.2.5 but it is in essence the dominant failure mode associated with *crack initiation and growth* due to a time-varying stress [40].

3. *Work Hardening* is a characteristic of ductile materials such as metals and alloys and occurs when there is overstress above the yield limit (σ_y). At low levels of work hardening it is not uncommon to see a shift in the residual stress state, leading to subtle changes in the performance of a device over time e.g. curvature or frequency of the device might shift. At higher levels this could eventually lead to catastrophic failure in the form of plastic deformation that completely degrades function of the device. Examples of susceptible components include LIGA MEMS, metal hinges, or contacts.

4. *Delamination*: Delamination between deposited layers has been observed in MEMS devices, and may occur (as shown in Fig. 3.19) either due to processing defects or high stress levels at interfaces. A processing defect occurs due to incomplete or non-uniform deposition and can be a source of weakness in a device, either leading to a non-performing part or failure of the device early in the life of the part. This can occur when the part is in the field as well and is more difficult to detect [41] because of the extent or location of the defect within the device. Typically this type of failure mode occurs due to a high stress event e.g. thermal or acceleration shock or anodic oxidation. Another form of delamination called *spall* occurs in multi-layered thin metal films which is a dynamic failure mode observed due to high strain rates in shock situations.

MEMS failures due to delamination may occur more often because of commonly used process steps such as wafer bonding, and chip-to-chip bonding (e.g. wafer-to-wafer bonding) which are very challenging to perfect. Results presented in

Fig. 3.19 Delaminated or debonded interface

[22], show that accelerated testing and thermal cycling can induce the onset of delamination, however the influence of mechanical shock on delamination is not as large as expected. Adhesion properties between different material layers or capillary forces could result in weakness of the bonded or interface layers; and thermal effects such as CTE mismatch between layers usually plays a very active role in causing such failures.

5. *Creep Failure*: Creep failures occur mostly in metals subject to a time dependent loading at an elevated temperature. There are several types of creep i.e. dislocation glide, Nabarro-Herring (NH), Coble creep, grain boundary sliding etc. and although a mechanism like dislocation glide occurs at any temperature, others like dislocation, NH or Coble creep typically occur when the homologous[10] temperatures (T_H) $T_H > 0.4$ and failures initiate at grain boundaries but it is also possible to activate secondary creep due to strain rate effects. The high stresses and gradients introduce time dependent behavior through dislocation glide and diffusion mechanisms and the strain levels can be large compared to the average size scale of a MEMS device. A good example of high stress states that might induce creep in MEMS is an RF switch that is in the on state, at an elevated temperature [42]. The primary concern is the use of certain metals as a structural material in MEMS because creep can occur even at room temperature degrading the performance of the device. Additional discussion on this topic and mechanism may be found in Section 4.2.4.

---

[10]Homologous temperature is the temperature of the material expressed as a fraction of its melting point using the Kelvin scale.

The dominant physical phenomenon (i.e. physics of failure) involved in each of these TM failure modes is due to material stress or strain beyond a certain limit.

3.2.2.2   Electrical (EL) Failures

Electrical failures occur due to static or dynamic charge transfer within materials or across gaps or surfaces and in MEMS devices this can lead to several potential failure modes.

1. *Dielectric Charging*: Dielectric charging is basically the accumulation of electric charge in an insulating dielectric layer. In certain MEMS applications like capacitive RF MEMS switches [39] the insulating dielectric between electrodes can accumulate charge over time leading to a failure where the switch will either remain stuck after removal of the actuation voltage or fail to contact under a sufficiently high voltage. In such switches, the mechanism for charge accumulation is a result of large electric fields across very thin dielectric layers [43]. The trapped charges have no conductive path and accumulate over time leading to two possible failure modes a) drift – the performance of the device changes slowly over time because of the stored charge, and b) latch-up – the accumulated charge changes the pull-in dynamics of the switch and can increase the pull-out voltage to a point where the mechanical restoring force is not sufficient to open the switch. Considerable effort has been devoted to both the experimental characterization of dielectric charging and the development of models that can be used to predict the impact of dielectric charging on electro-mechanical behavior of a capacitive switch [18]. Further discussion is available in Section 4.3.1.
2. *Electromigration*: In semiconductor devices, *electromigration* is a well documented phenomenon caused by the formation of voids or hillocks that may occur over time, due to high current densities in thin-film conductors within integrated circuits. Figure 3.20 shows a damaged interconnect due to significant momentum

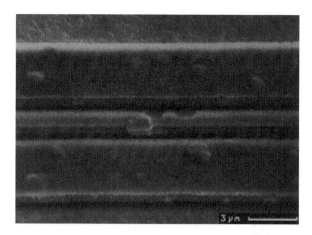

Fig. 3.20 SEM Micrograph showing voids and hillocks (reproduced with permission Copyright – [44])

transfer from electrons to conductor atoms [44]. The failure modes with this phe-nomenon are quite clear – loss of function due to shorts, and change in parasitic impedances over time. In the design phase, Black's empirical equation to pre-dict MTTF of a wire, factoring in electromigration may be used to estimate the effects of current density and temperature on reliability. Section 4.3.3 delves into more detail of this in-use failure mode.

3. *Electro-static Discharge (ESD) or Arcing*: Another potential electrical failure mode that commonly results in catastrophic damage in MEMS is ESD or arc-ing (Fig. 3.21). The presence of very small gaps (order of a few nanometers to a few microns) with the possibility of gap closure (during use) and the geome-tries that can lead to non-uniform high electric field, make MEMS structures particularly vulnerable to electrostatic discharges (ESD), overvoltage, charging, or corona effects. There is limited research in this area [45] and more work is needed particularly in the area of RF MEMS switches.

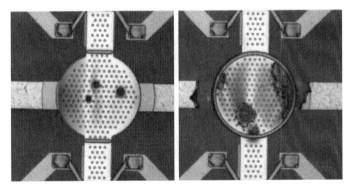

Fig. 3.21 Catastrophic failure due to ESD in RF MEMS switches (reprinted with permission Copyright – 2008 IACM, ECCOMAS [45])

The reliability of MEMS devices can be heavily influenced by electrical failures and it is important for the designer to understand the sources of these failures and account for them. A more detailed treatment of this topic is available in the next chapter (Section 4.3.2).

3.2.2.3 Environmental (ENV) Failures

MEMS applications are diverse in their interaction with the environment. In some applications, such as pressure sensors and microphones the sensing element is directly exposed to the operating environment which could in some cases be quite aggressive. In a harsh environment application such as tire pressure monitoring (TPMS), the sensor element has to be able to directly sense the air pressure on one side of the diaphragm, however besides the application stresses and temperatures, this environment typically contains particulates of various sizes, and a multitude of contaminants. The interactions between environmental forces and the materials within the device can result in a variety of failure modes in MEMS. Some of these are listed below and is covered in more detail in Chapter 4.

1. Anodic Oxidation: Anodic oxidation can be a fatal failure mechanism in polysilicon MEMS devices that operate in humid environments. The exact failure mode depends on the design but it is known that positively charged polysilicon traces can fail due to oxidation [46], and that such oxidation can cause delamination between polysilicon and nitride layers [41]. A more thorough discussion of anodic oxidation is available in Section 4.4.2.
2. Corrosion – There are many types of failure mechanisms related to corrosion including galvanic, crevice, pitting corrosion, stress corrosion, dendrite growth, whisker growth, and corrosion due to moisture, microorganisms and biological contamination. While it is not possible to cover in detail all these types of corrosion the reader is directed to Section 4.4.3 for more information on galvanic corrosion.
3. Grain growth: Grain growth in MEMS materials like polysilicon has been observed under conditions of high stress and temperature leading to failure mechanisms where the device ceases to function or there is a gradual change in behavior. From experiments on polysilicon [47] it is has been observed that grain growth mechanism is significantly affected by the doping conditions specifically when the dopant concentration in the grains is above the solid solubility limit (and is apparently independent of the method of polysilicon doping). Generally, polysilicon grows by secondary recrystallization which is driven by grain boundary energy as opposed to defect energy and the rate is temperature and stress dependent. In particular geometrical features such that cause higher stress states (e.g. film corners) will have lower grain growth.

In the Chapter 4, a more detailed look at other environmental failure mechanisms caused by radiation as well as the physics of failure involved in selected cases is presented but for now we will continue to look at failure modes in MEMS that have their origin in the product development phase.

3.2.3 Non-analyzed Conditions

Non-analyzed conditions is a sub-category of failure modes that basically is a catch-all for the many different environmental (or other) conditions the device may be subjected to that are not analyzed *a priori*. It is impossible to simulate all possible environmental or operating conditions prior to fabrication, and so a robust strategy of testing and qualifying the device under a series of burn-in, acceleration and other protocols is used to reveal weaknesses or uncover potential failure modes. Examples of these are described in Chapter 2.

A good example of such a factor is the stress corrosion cracking of polysilicon [48]. In the absence of a corrosive environment, a brittle material like polysilicon

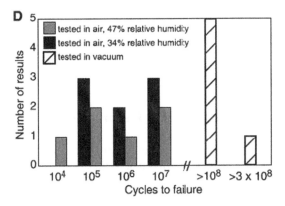

Fig. 3.22 High cycle fatigue of polysilicon (reprinted with permission Copyright – 2002 Science [48])

should be relatively insensitive to cyclic fatigue but such fatigue effects are observed in MEMS polysilicon samples tested in air [49] as seen in Fig. 3.22. The fatigue damage may originate from contact stresses at surface asperities; which exacerbates subcritical crack growth during further cyclic loading.

Under these conditions, a corrosive ambient such as laboratory air exacerbates the fatigue process through formation of an additional thickness of surface oxide on surface asperities or crack surfaces which generates higher stresses during compressive stress cycles. Without cyclic loading, polysilicon does not undergo stress corrosion cracking.

3.2.3.1 Leakage Currents

Leakage currents in MEMS can cause havoc in the performance of the part over life. Previous studies [50] have researched the implications of leakage currents in surfaces and volumes of dielectrics within the MEMS device itself but quite often the failures due to leakage currents are not limited to the sensor itself but could occur due to assembly or semiconductor processing.

Most MEMS have some control circuitry and I/O pads that influence performance of the part, and often it is here that we encounter leakage currents that can cause reliability failures. The use of conductive die-attach or even silicon chip outs from wafer dicing or handling can lodge in bond pad regions or exposed interfaces leading to leakage currents between isolated parts of the design. The presence of an oxide layer on the surfaces of such particles makes it less likely that these will be detected at final test and during field use but when the oxide starts to degrade due to stress effects, the resulting leakage currents could start to influence part performance. In traditional IC chips, such failures are detected by techniques such as XIVA/LIVA/CIV[11] and similar techniques can be used for MEMS.

---

[11]XIVA – External Induced Voltage Alteration, LIVA – Light Induced Voltage Alteration, CIVA – Charge Induced Voltage Alteration.

In summary, the functional and material failure modes described in this section describe the major failure modes encountered in MEMS design. In the next section we will look at failure modes that have their origin in the manufacturing phase of product development.

3.3 Manufacturing Failure Modes

Manufacturing related failures are due to specific processing characteristics and are usually difficult to eliminate – they are mainly of two types depending on where they originate in the manufacturing process. In general, the manufacturing process is divided between front-end processing which typically includes specific clean room processing, photolithography, etching etc., and back-end processing which includes wafer dicing, assembly and final test. Figure 3.23 identifies some of the main manufacturing related steps where defects may be introduced. Sometimes it is also common to use the terms *local* and *global* defects rather than front end or back end. *Local* refers primarily to contamination and any form of mis-processing such as a voids or stringers, and could potentially also include effects of design rule sensitivity; whereas global defects include a broader spectrum of defects from those due to wafer level variations and handling, to assembly.

Fig. 3.23 Manufacturing process related defects

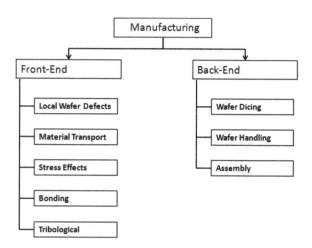

3.3.1 Front End Process Defects

Process related defects fall into three broad categories – *material transport* failures such as those due to deposition and etch steps, *wafer bonding* failures such as hermiticity, and *tribological failures* such as stiction.

3.3.1.1 Local (Wafer) Defects

Local defects due to contamination are a common failure mode that is routinely dealt with throughout the industry. In general, local defects in MEMS are primarily of four types – *particulate, ionic, organic* contamination defects, and *voids* and *stringers.*

Particulate defects refer to nano or micron size particulates (typically FEOL[12] 18–90 nm and 0.052–0.064 #/cm^2 and BEOL[13] 36–180 nm; 0.052–0.064 #/cm^2 [51]) that will cause a variety of potential failure modes in ASICs but in MEMS larger particles can be tolerated. The small feature size (Fig. 3.24) especially if conductive, could potentially cause open/short circuits or degrade motion of the MEMS element or even cause intermittent performance deviations (Fig. 3.25).

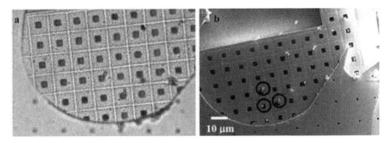

Fig. 3.24 Particulate contamination on a passive shock element (reprinted with permission Copyright – 2008 Sandia [52])

Obviously, the size of the particle has much to do with the physical kinetics of the particle in terms of what forces activate them, and more importantly on what can be done to eliminate particles below or above a certain threshold size. For example, for particles <10 nm in size; the particle motion is dominated by *Brownian* motion and is heavily influenced by gas or liquid molecules. Typically particles that are created close to the wafer surfaces may be deposited due to Brownian motion. At the next size up – between 0.1 and 1 μm, the motion of the particle is influenced by *thermophoresis* which is a non-continuum effect caused by the temperature gradient e.g. cold wafer introduced into a hot oven. In fact, particles are repelled from hot surfaces and attracted to cold surfaces leading to higher contamination levels in these cold regions [53]. Lastly, at sizes above a critical size d_{cr} (1 μm or greater) inertial or gravitational forces will dominate.

The adhesion and removal of particles from wafer surfaces [54] is of critical importance to MEMS manufacturing because these particles can directly result in defects leading to failure. Adhesion forces are categorized [55] as follows:

---

[12] FEOL – Front-End-Of-Line.
[13] BEOL – Back-End-Of-Line.

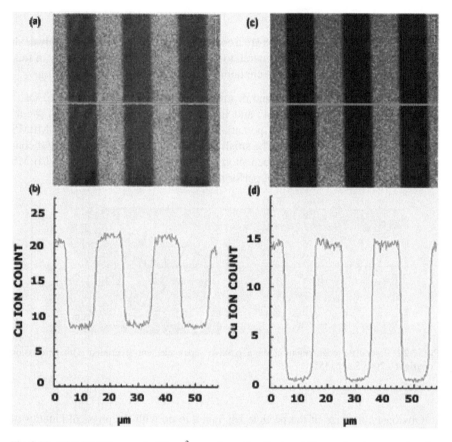

Fig. 3.25 TOF-SIMS image of a 60 μm² area of a wafer after two different cleaning steps showing copper ion counts along the corresponding highlighted lines (reprinted with permission Copyright – 1999 Micromagazine.com [57])

- Adhesion forces that dominate in the region of the contact and immediate surroundings such as *Van der Waals* and *electrostatic* forces.
- Adhesion forces due to chemical bonding such as *EDL – electrical double layer* (function of solution pH) or *Hydrogen and Covalent Bonds* (e.g. SiO_2/Glass). An EDL forms when particles in solution become charged and the *zeta potential*[14] affects particle deposition.
- Adhesion forces caused by interfacial reactions such as *diffusion, condensation* or *diffusive mixing* (RH dependent)

---

[14] Zeta Potential – Potential at shear plane.

Once a particle is in contact with the surface it is not uncommon for adhesion induced deformation [56] to occur which increases with time resulting in a decrease in removal efficiency of such particles.

Ionic Defects generally refer to the presence of metallic or non-metallic ions on the surface of the wafer. The metallic ions are highly mobile and can cause charging of dielectrics or other layers which can directly impact performance. However, more problematic is the presence of ionic defects in the presence of strong local electric fields which may tend to concentrate the accumulation of ions on the surface of the wafer. Ions can be detected with a variety of techniques (discussed in Chapter 5) but TOF-SIMs (Fig. 3.25) and TXRF are quite common [57]. A possible technique to mitigate such defects before field testing is *burn-in*, although a wash followed by burn-in may be more effective (Fig. 3.26).

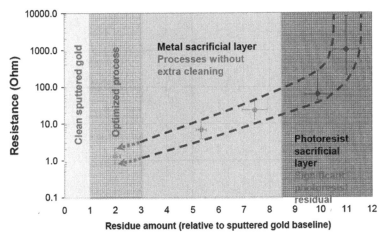

Fig. 3.26 Switch contact resistance as a function of organic residual contamination level determined by Auger spectroscopy (reprinted with permission Copyright – 2007 SPIE [36])

Organic Defects are generally carbon based organic solvents used in fabrication processes and comprise of typical molecules from sources such as – photoresist (polyimides, SU-8 etc.), methyl alcohol, acetone, isopropyl alcohol etc. that are due to wet or dry wafer cleaning operations. The effect of such organic defects can be generally small during normal operation of the device but occasionally there might be situations where they could lead to failures [36]. A good example is the accumulation of organic molecules at sites with high field fluctuations, or the contribution of organics to degradation of anti-stiction coatings. Optimization of the cleaning process is necessary to minimize the presence of organic contaminants at the end of wafer fab processing. One method for removal of these defects from critical surfaces before field testing is burn-in (48 h at 150°C in dry atmosphere) but it is generally better to perform cleaning steps prior to hermetic sealing or packaging to minimize the possibility of field failures.

Voids and Stringers are isolated defects that have been observed in MEMS processing either due to chemical or physical conditions that the wafer is subjected to during fabrication. Voids have been observed to form in a variety of process conditions including deposition, annealing, corrosion, etc. For example, stress induced voiding is a common occurrence in IC manufacturing (Fig. 3.20), and is observed in trace wires usually due to electron migration at grain boundaries. Stringers or streamers on the other hand, are due to incomplete process steps or a marginal violation of a design rule, which gives rise to residual stringers within the moving MEMS element. These can linger even after cleaning steps and may subsequently lead to failures in the field. The example of a stringer shown in Fig. 3.27 was detected after a short-circuit was detected during operation of the device. The failure analysis revealed that the stringer created an electrical path between two conductive adjacent surfaces.

Fig. 3.27 Example of a stringer lodged in an isolation trench (reprinted with permission Copyright – 2000 Coventor)

Local wafer defects such as those described in the previous section are very common in most IC manufacturing but in MEMS manufacturing these same defects cause a variety of reliability issues in the field.

3.3.1.2 Material Transport – Deposit/Etch Failures

A variety of process deposition and etch defects are routinely encountered in MEMS fabrication.[15] In modern IC and MEMS manufacturing there are established techniques [12] to identify defects and discard the specific die where these defects occur. The presence of defects is more often than not a yield issue where the die or part will simply not function as intended or at all and since this can typically be detected at final test it does not specifically pose a problem for long term reliability but there

---

[15] For details of MEMS fabrication processing the reader is referred to (2).

are certain classes of defects that will not be detectable by electrical testing and only manifest themselves in the field. It is these defects that cause reliability issues and lead to degraded performance of the part.

In MEMS fabrication there is a wide variety of material addition techniques and from time-to-time deposition defects will occur in all of them either because of the tool used or because of interaction of the design and the flow conditions within the process zone. While it is impossible to go into great detail with each process step used in the MEMS industry we will look briefly into a few such steps and the defects they produce to give the reader a basic idea of which failure mechanisms can impact overall reliability.

- Chemical Vapor Deposition (CVD): The primary defect classes that occur during this deposition step are point defects, clusters, dislocation, and stacking faults. Due to the high temperatures (above 600°C) and relatively low deposition rates some of these defects can form weakness (crack initiation sites) or grow larger (grain size structures) that could lead to anomalies. CVD results in the conformal deposition of material over the previously deposited layer and this creates some potential vulnerable areas in the design, specifically like anchor locations and steps where defects can lead to reliability issues.
- Photolithography: There are several types of photolithography related defects that can be generated during fabrication. Defects that prevent motion of the MEMS element are the most easily detectable because of self-test (or BIST[16]) or final electrical test because a defect such as a particle or residue that obstructs the motion of a proof mass or finger will not respond to the applied stimulus in the same way and can thus be effectively screened out. There are several sources of photo-track-induced defects [12] such as bubbles in the developer dispense, incomplete post-develop rinsing, scumming of the resist etc. and each will interact with a given design in a unique way leading to different defect size and localization distributions.

  Reliability issues are also caused by the same defects when they are undetectable by electrical screening. In this case, defects like particles that are smaller than a critical gap, or residue that adheres to the moving element can be undetectable during the electrical test. Optical inspection on each and every die is prohibitively time consuming and expensive for high volume applications and is not really an effective solution. Quite often the development of an electrical test to screen out offending die is the only path for corrective action but ultimately optimization of the process to remove these defects is necessary.
- Evaporation and Sputtering are similar processes for depositing materials on the surface of a MEMS wafer. The primary defects are point defects that form either due to the characteristics of the process tool or because of a pre-existing

---

[16] BIST – Built-In Self Test.

local defect (e.g. particles). For example, in the evaporation of gold using an e-beam evaporator one can sometimes observe local defects shaped like round balls (diameter ~100 nm or larger) known as "spit" gold which are contaminants on the target surface that act as a catalyst. It is possible for vibration to dislodge these defects resulting in particulate defects during field operation leading to a reliability failure.

There are several other common deposition techniques such as electrodeposition, and thermal oxidation that also can result in localized defects.

Etch related failures are also fairly common sources of defects that result in reliability failures. An example of a particularly common problem is the variation of etch characteristics across a wafer. As can be seen in Fig. 3.28, the thickness of deposited nitride across the wafer can vary significantly at the edges of the wafer compared to the center. This typically results in parts with a variety of different responses and it is entirely possible to get parts that are close to set limits and even failures. A more severe problem is that of marginal parts, which are nothing but parts that are within acceptable performance limits but close to the margins. In operation, these parts can quickly fail due to particular in-field stresses (Fig. 3.29).

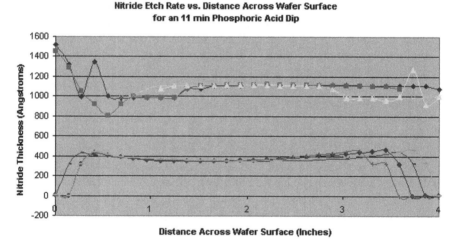

Fig. 3.28   Process etch variation across the wafer

Finally, one should remember that process steps involved in MEMS just like those in semiconductor processing are highly controlled chemical reactions that can also produce failures due to incompleteness of the reaction. For example, a resist removal step can be incomplete due to insufficient process time, physio-chemical differences in material, etch design rules, etc. The material left behind can then cause failures in the form of the above described *Local Defects*.

Fig. 3.29 Stress relaxation behavior for pure Al films (<100 nm) as a function of temperature (reprinted with permission Copyright – 2000 Applied Physics Letters [60])

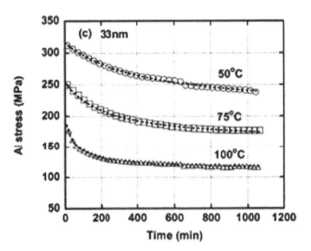

3.3.1.3 Stress Relaxation Effects

There are several types of thin-film stress effects that are commonly encountered during MEMS fabrication that can influence the long-term reliability of a MEMS device. The effects of residual stress, creep, and fracture were discussed earlier in this chapter[17] but it is important to mention the effects of stress relaxation effects brought about by fabrication conditions and which can impact long term reliability.

The investigation of stress relaxation in nanoscale thin films (such as aluminum – [58]) have found that the relaxation is strongly dependent on temperature and film thickness, with the relaxation rate being highest for the highest temperature and the thinnest films. In polysilicon, stress relaxation is negligible at room temperatures but has been observed at elevated temperatures above 1000°C [59]. In metals however, the relaxation mechanism is attributable to dislocation motion or grain boundary sliding [60] and in metals annealing is commonly used to relax metal stress at relatively low temperatures.

The relaxation of metallic thin films can adversely affect the performance of a MEMS device in many different ways. The use of metal films in coatings (for optical devices) or conductor electrodes (RF switches) makes it important to factor in the stress relaxation into the design. An alternate approach is to effect the relaxation through an annealing step (at elevated temperature) during manufacture.

3.3.1.4 Process Tribological Failures – Stiction

Stiction is one of the primary tribological failure mechanisms in MEMS, and occurs where suspended structures are pinned unexpectedly due to adhesion which might occur during contact of proximal surfaces [61, 62]. In MEMS, particularly surface

---

[17] Additional reference is Chapter 4.

micromachined structures, the surface area to volume ratio is large, and the stiffness of restoring springs is typically small, which makes these proximal surfaces particularly prone to stiction which may occur during processing or in the presence of liquid (e.g. elevated RH levels) [63]. Stiction as a phenomenon can also occur during use (e.g. shock), and is occasionally called *in-use* stiction, as it dependent on the state of the surfaces and specifically the surface energies post manufacture.

Stiction as a failure mechanism can be understood by considering the adhesion of the two surfaces in contact. Adhesion occurs either due to van der Waals, electrostatic forces (trapped charge), capillary forces or a combination thereof [66]. During fabrication, the use of wet chemical processes can leave behind ions and dangling bonds as well as minute amounts of water from trapped liquid due to pressure differences and surfaces tension forces. When the surfaces of two solids are brought close to each other (solid-solid contact), a surface force arises due to direct interaction between the molecules or atoms at the surfaces [32], and this force can be positive or negative depending on the proximity of the surface pair. In liquid mediated contact, the adhesion arises due to surface tension forces and this adhesion energy can be quantitatively measured using analytical models [67] and test structures [68, 69], such as the free-standing cantilever beams shown in Fig. 3.30.

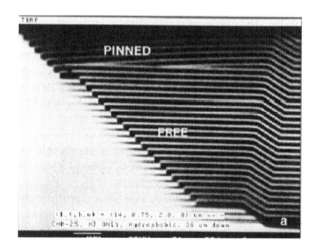

Fig. 3.30 Pinned and free standing cantilever beams – Reprinted with permission Copyright – 1993 IEEE [64, 65]

In MEMS devices, the contact between surfaces can occur horizontally or vertically depending on the particular design. The electrostatic force between surfaces usually has to be factored into the force balance as shown in the diagram in Fig. 3.31:

The electrostatic pull-in force brings the two contacting surfaces together but it is the restoring force that has to overcome stiction if the device is to function correctly [70]. In most devices, the restoring force is typically enabled through a spring-like structure with a constant restoring force or a bias in the opposite direction.

The surface energy U_s is simply defined in terms of the contact area A_c and the work per unit area (2γ) required to separate the surfaces to ∞.

Fig. 3.31 Pull-in and pull-out curve for a typical MEMS micromachined structure

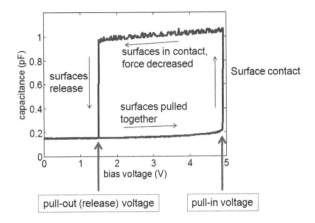

Fig. 3.32 Cantilever beam adhering to the substrate in an (**a**) S-Shape and (**b**) arc shape – Reprinted with permission Copyright – 1993 IEEE [64]

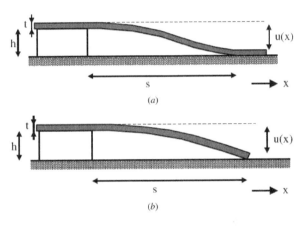

$$U_{\mathrm{s}} = 2A_{\mathrm{c}}\gamma \qquad (3.4)$$

If we consider a *Hertzian*[18] contact with circular contact area (πa^2), the total contact force is the sum of the mechanical force and the adhesion force which is given by the following expression:

$$F_{\mathrm{adhesion}} = 4\sqrt{\gamma E A_{\mathrm{c}}} \qquad (3.5)$$

For non-circular contact areas the adhesion force is determined by using the same equations with an equivalent radius (*a*). The surface energy γ is highly dependent

---

[18] *Hertzian* contact stress refers to the localized stresses that develop as two curved surfaces come in contact and deform slightly under the imposed loads.

on processing conditions and can be modulated with the use of anti-stiction coatings (e.g. SAMs[19]) or particular drying or cleaning steps [66, 71].

The simplest strategy employed to quantify stiction in a given process is by measuring the free standing lengths of cantilevers fabricated in the same process. For simple cantilevers, derived formulae [64, 65] which relate the contact length to the beam stiffness and adhesion energy for both simple configurations of a deformed cantilever, i.e. "S" and "Arc" shape (Fig. 3.32) provide the relationship between the dimensions of the beam, and surface adhesion energy. Below some limit the beam will not stick to the substrate and to determine the surface energy for a material and process one can use this technique to measure the detachment length from an array of cantilevers and fit the values to equation (3.6). As the free standing length of the beam approaches the beam length, the cantilever pivots and changes from an S-Shape to an Arc-Shape. The models assume that adhesion energy *includes* elastic deformation of the substrate

and the surface energy γ_s is given by:

$$\gamma_s = \frac{3}{2} \frac{Et^3h^2}{(l-l_s)^4} \tag{3.6}$$

The physical state of the contact surfaces is a critical factor in the existence and development of stiction during use. In simple terms, a high surface energy will make it easier for stiction to play a role in the performance of the device, and since surface chemistry is easily affected by a variety of factors such as oxide or contaminant films, moisture, roughness, ambient gas, and obviously the design, it becomes necessary to consider the balance and control of these factors in preventing stiction [66]. It is also now easier to understand how the surface energy may change over time either due to repeated contact (from pull-in or shock) or due to a change in moisture levels, or a change in ambient gas or surface roughness which could result in higher surface energies and stiction during field operation.

3.3.1.5 Wafer Bonding (or Hermiticity)

A significant number of MEMS devices produced today are hermetically sealed using wafer-wafer bonding which have enabled hermetic packaging of MEMS die before they leave the fab line so as to minimize contaminations from particles and ambient gases. In several applications, the MEMS element needs a lower pressure or vacuum conditions to perform optimally. In these cases the hermitic seal is created by traditional wafer bonding methods which include anodic, glass frit, eutectic, solder, reactive and fusion bonding [72]. It is also common to see a wide variety of gases or gas mixtures (N_2, Ne, etc.) employed for optimal performance of the MEMS sensor element. Other techniques such as anodic bonding and solder

[19] SAM – Self-assembled monolayer.

bonding are known to have issues such as a lower limit for ambient pressure and contamination because of generated or surface desorbed gases.

In a majority of MEMS sensors today, glass frit bonding is by far the most common technique for achieving a hermetic seal. The glass is composed of solvents, organic binder, lead borosilicate glass, and alumina silicate glass (cordierite) and is screen printed onto one wafer, dried and then the temperature is increased until the glass melts (glaze) above 400°C. The second wafer is then aligned and bonded under pressure to the first wafer, followed by slowly cooling the wafers back to room temperature. There are several failure mechanisms that can be traced back to the bonding process and the common ones are incomplete seal glass coverage (Fig. 3.33), squish out of the seal glass on both sides of the seal, lead particles, glass cracks, incomplete adhesion, etc.

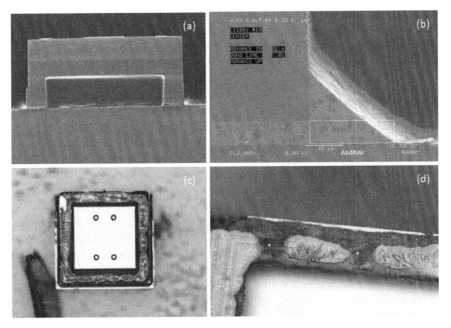

Fig. 3.33 Cross-section of a bonded sensor showing (**a**) cross-section, (**b**) seal glass squish out, (**c**) incomplete seal glass coverage and (**d**) gaps in seal glass (reprinted with permission Copyright – 2008 Analog Devices)

In devices packaged at atmospheric pressures any loss of hermiticity due to the above mentioned failure modes, makes the device susceptible to the ingress of undesirable gases or elements from the ambient operating environment. The external packaging of the device can also make a significant difference. In the case of plastic over-molded packages, the diffusion rate of moisture through the package is relatively quick and so moisture can enter the finest of gaps (through capillary action) in the glass seal and over time can accumulate within the cavity causing stiction or even corrosion. One noteworthy point is that these types of failures maybe undetectable before failure because the rates of moisture ingress could be very slow.

In the case of cavities sealed below or above atmospheric pressure, the failure of the seal glass to maintain hermiticity will cause the cavity pressure to immediately revert to atmospheric pressure, resulting in a measurable change in performance of the device. In this case, a manufacturing defect such as incomplete seal glass coverage could be detected before the part is in the field. However, in the field a seal glass failure could occur due to external loading conditions like shock and the resulting performance change could be detrimental to the intended application.

3.3.2 Back End Process Failures

In MEMS product development, back end process steps include all the steps after the final wafer fabrication step such as dicing, assembly process steps, and ATE[20] testing. These steps can introduce a variety of different failure modes and we will limit our discussion to just a few of the worst offenders and will highlight others.

3.3.2.1 Wafer Dicing

Wafer dicing is the process step where the MEMS wafer is sent to a high speed saw that is capable of cutting the wafer along predefined streets to singulate individual dice. Dicing is typically done with a diamond tipped blade few *mils*[21] thick. Silicon is a very brittle material and during a high-speed mechanical sawing operation, blade vibration in the presence of diamond particles and coolant can cause damage to sensitive die. A majority of these particles can be removed with subsequent cleaning steps but a fraction can linger in reentrant gaps and grooves in the die leading to failures due to silicon chips or chip-outs that break away from diced surfaces as shown below in Fig. 3.34. Visual inspections during this dicing step and careful cleaning after are necessary to minimize the proliferation of such chip-outs.

The presence of chip-outs within the final package can degrade the performance of the part or cause catastrophic failure which may not be detected during the final testing of the part leading to field escapes where the prevalence of certain conditions can lead to part failure.

Other dicing techniques such as cleaving and stealth dicing are also commonly used but these depend on the thickness and crystal orientation of the wafer as well as cost. Lastly, DBG (Dice Before Grind) is a singulation process primarily developed for separating dice and is employed when normal sawing would created unacceptable levels of chipping and edge damage. This technique has been demonstrated for ultra-thin die as thin as 25 μm [73].

---

[20] ATE – automatic test equipment.

[21] Mil – a thousandth of an inch.

Fig. 3.34 Evidence of chipping from the cut surface during a dicing step

3.3.2.2 Wafer Handling

Wafer handling is another important BE process step that can introduce failure modes which can compromise the reliability of the part. Wafer handling occurs both during front-end and back-end operations but it is during back-end operations that handling of the wafer becomes more sensitive because of the difference in operating environments between front and back end lines.

A common example is shown in Fig. 3.35 above where the level of training of fab technicians is directly correlated to the number of scratches introduced on the surface of a wafer. Damage related to scratches can be a major reliability issue depending on the location and severity of the scratch. The scratch can easily become an initiation site for a crack that propagates from the top or bottom surface of the wafer through the thickness, and the resulting defect may not cause failure immediately but result in a weakened part that then becomes a reliability issue.

Fig. 3.35 Differences in number of scratches per wafer between trained and untrained technicians

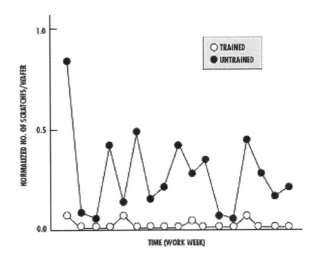

3.3.2.3 Packaging

In Section 3.2.1.3 we discussed the major design related failure modes encountered in packaging and assembly. The field of electronic packaging is broad and the reader can find plenty of references on specific failure modes pertaining to the manufacture of different kinds of packages, as well as wire bonding, die-attach and over-molding [74]. From a MEMS manufacturing perspective there are two important areas related to the manufacturing of the part that can significantly influence the part reliability – package manufacturing design rules, and material control.

Assembly design rules used to create a MEMS packaged part are usually identical or at least very similar to those used in the packaging of IC chips. For example, in the case of a two-chip packaged part, where the MEMS and IC chip co-exist in the same package, the edge of a MEMS die has to be placed a certain minimum distance away from the edge of the ASIC pad row to allow for wire bonding. This design rule exists to prevent *squish-out* of the die-attach from contaminating the pad row and allow for proper wire bonding, since it is possible that the design rule could be violated due to a combination of factors including die placement tolerance, die-attach cure conditions, or other assembly process conditions. Strict incoming and process quality screening for contaminants/foreign materials as well as design rule violations is absolutely necessary to minimize the risk of internal corrosion in MEMS products.

Material control involves optimization of the assembly materials set (and processes) to achieve certain repeatability in the package construction which is very intimately connected to the MEMS sensor. There are many examples to chose from but in packaging MEMS devices, the stress state of the device is of critical importance and as we have seen before, there are many factors that can modify the device performance. In plastic over-molded packages, for example, the effect of post-mold cure time influences mechanical strength, glass transition, and adhesion strength and may play a significant role in the reliability of the part.

In summary, assembly processes may quantitatively influence the overall reliability of the part. The process controls in the back-end need to be as stringent as those in the front end to ensure high reliability of the overall product.

3.4 Summary

In this chapter, we have looked at a variety of MEMS failure modes that have their origin in the design or manufacturing phases of the product development cycle. A fair majority of the design failure modes can be avoided with robust design practices that are usually cumulative in the sense that over time the design team adds analyses that can predict the propensity of a part to cause yield or reliability issues. In terms of manufacturing failures modes, the ability to avoid failure modes that can impact part reliability is more tenuous because of two main factors – the sensitivity of front end processes to MEMS device performance, and the interaction of the MEMS device with its immediate surroundings. The control of these factors has to be very good to avoid failure modes that can impact part reliability. In the next chapter, we will look at the mechanisms of certain more common MEMS in-use failure modes in more detail.

References

1. Pineda de Gyvez, J., Pradhan, D. (1999) *Integrated Circuit Manufacturability – The Art of Process and Design Integration*. Piscataway, NJ: Standard Publishers Distributors IEEE Press.
2. Madou, M.J. (2000) *Fundamentals of Microfabrication*. Boca Raton, FL: CRC Press (Graduate level introduction to microfabrication. Fifth printing, Fall 2000).
3. Milek, J.T. (1971) *Silicon Nitride for Microelectronics Applications*. New York: IFI/Plenum.
4. Peterson, K.E. (May 1982) Silicon as a mechanical material. Proc. IEEE. 70(5), 420–457.
5. Sze, S.M. (1994) *Semiconductor Sensors*. New York: Wiley Inter-Science.
6. See-Ho Tsang, A.P. (2005) Mumps Design Book. [Online] April 10, 2010. www.memscap.com.
7. Amerasekera, E.A., Najm, F.N. (1997) *Failure Mechanisms in Semiconductor Devices*, 2nd edn. New York: Wiley.
8. Mariani, S. et al. (2009) Modeling impact-induced failure of polysilicon MEMS: A multi-scale approach. Sensors. 9, 556–567.
9. Srikar, V.T., Senturia, S.D. (June 2002) The reliability of microelectromechanical systems (MEMS) in shock environments. J. Microelectromech. Syst. 11, 206–214. ISSN 1057-7157.
10. Senturia, S.D. (2001) *Microsystem Design*. Boston, MA: Kluwer.
11. Younis, M., Miles, R., Jordy, D. (2006) Investigation of the response of microstructures under the combined effect of mechanical shock and electrostatic forces. Micromech. Microeng. 16, 2463–2474.
12. Grosjean, D.E. (2003) Reducing defects in integrated surface-micromachined accelerometers. micromagazine.com. http://www.micromagazine.com/archive/03/03/grosjean.html.
13. Goering, R. (2005) EE Times. [Online] March 10, 2010. http://www.eetimes.com/showArticle.jhtml?articleID=173602466.
14. da Silva, M.G., Bouwstra, S. (2007) Critical comparison of metrology techniques for MEMS. Proceedings – SPIE The Internation Society for Optical Engineering. San Jose, CA: Internation Society for Optical Engineering, p. 646.
15. Brown, S.B., Van Arsdell, W., Muhlstein, C.L. (1997) Materials reliability in MEMS devices. Proceedings of International Conference on Solid-State Sensors and Actuators, Chicago, pp. 591–593.
16. Van Kuijk, J., Schropfer, G., da Silva, M. (2005) Design automation for MEMS/MST. G. Design, Automation and Test in Europe Conference & Exhibition, Munich.
17. Coventor. (2009) MEMS design and analysis manual (Tutorial – Vol 1). Cary, NC: Coventor Inc., pp. T4–94.
18. Yuan, X. et al. (2006) Temperature acceleration of dielectric charging in RF MEMS capacitive switches. IEEE Trans. Devices Mater. Reliab, 6(4), 556–563.
19. Douglass, M.R. (2003) DMD reliability: A MEMS success story. In D. Tanner, R. Ramesham (ed) *Reliability, Testing and Characterization of MEMS/MOEMS II*. Proc. Of SPIE Vol. 4980, SPIE, Bellingham, pp. 1–11.
20. Tummala, R.R., Pymaszewski, E.J., Klopfenstein, A.G. (1998) *Microelectronics Packaging Handbook*. New York: Chapman & Hall. 0-412-08561-5.
21. Hsu, T.R. (2006) Reliability in MEMS packaging. 44th International Reliability Physics Symposium, San Jose, CA, March 2006.
22. Swaminathan, R. et al. (ed) (May 2003) Reliability assessment of delamination in Chip-to-Chip bonded MEMS packaging. IEEE Trans. Adv. Packaging. 26(2), 141–151.

23. Felton, L.E. et al. (2004) Chip scale packaging of a MEMS accelerometer. Proceedings of the Electronic Component and Technology Conference, Florida, pp. 870–873.
24. Dixon-Warren, St.J. (2009) Chipworks looks inside the freescale harmems process. MEMSBlog – The MEMS Industry Group Blog. [Online] http://memsblog. wordpress.com/2009/08/03/chipworks-looks-inside-the-freescale-harmems-process/.
25. Joo, J.W., Choa, S.H. (2007) Deformation behavior of MEMS gyroscope sensor package subjected to temperature change. IEEE Trans. Compon. Packag. Technol. 30(2), 346–354.
26. Tada, K., Fujioka, H. (1999) Properties of molding compounds to improve package reliability of SMDs. IEEE Trans. Compon. Packag. Technol. 22(4), 534–540. ISSN: 1521-3331.
27. McNeil, A.C. (1998) A parametric method for linking MEMS package and device models. IEEE, Proceedings of the Solid-State Sensors and Actuators Workshop, Hilton Head, SC, pp. 166–169.
28. Bart, S.F. et al. (1998) Coupled Package-Device Modeling for MEMS.
29. Stellrecht, E., Han, B., Pecht, M.G. (2004) Characterization of Hygroscopic swelling behavior of mold compounds and plastic packages. IEEE Trans. Compon. Packag. Technol. 27(3), 499–507.
30. Walwadkar, S.S., Cho, J. (2006) Evaluation of die stress in MEMS packaging: Experimental and theoretical approaches. IEEE Trans. Compon. Packag. Technol. 29(4), 735–743.
31. Rebeiz, G.M. (2003) *RF MEMS Theory, Design and Technology*. Hoboken, NJ: Wiley. 0-471-20169-3.
32. Bhushan, B. (2002) *Introduction to Tribology*. New York: Wiley.
33. Becher, D. et al. (2002) Reliability study of low-voltage RF MEMS switches. Proceedings of the GaAs MANTECH Conference. San Diego: Mantech, p. 54.
34. Hilbert, J.L., Morris, A. (2002) RF MEMS Switch platforms expedite MEMS Integration and Commercialization. COMS 2002, Ypsilanti, MI.
35. ASME. (2003) *Course 469: MEMS Reliability Short Course*. s.l.: ASME.
36. Ma, Q. et al. (2007) Metal contact reliability of RF MEMS switches. *Reliability, Packaging, Testing, and Characterization of MEMS/MOEMS VI*, Vol. 6463. San Jose, CA: SPIE.
37. Mihailovich, R.E. et al. (2001) MEM relay for reconfigurable RF circuits. IEEE Microw. Wirel. Compon. Lett. 11, 53–55.
38. Hyman, D. et al. (1999) Surfacemicromachined RF MEMS switches on GaAs substrates. Int. J. RF & Microw. Comput.-Aided Eng. 8, 348–361.
39. Goldsmith, C. et al. (2001) Lifetime characterizat ion of capacitive RF MEMS switches. IEEE MTT-S 2001 International Microwave Symposium Digest. pp. 227–230.
40. Suresh, S. (1991) *Fatigue of Materials*. Cambridge: Cambridge University Press.
41. Plass, R. et al. (2003) Anodic oxidation-induced delamination of the SUMMiT^{TM} Poly 0 to silicon nitride interface. Proceedings of SPIE, Vol. 4980, pp. 81–86.
42. van Gils, M., Bielen, J., McDonald, G. (2007) Evaluation of creep in RF MEMS devices. *Thermal, Mechanical and Multi-Physics Simulation Experiments in Microelectronics and Micro-Systems*. Nijmegen: NXP Semicond, pp. 1–6. 1-4244-1106-8.
43. Sumant, P.S., Cangellaris, A.C., Aluru, N.R. (2007) Modeling of dielectric charging. Microw. Opt. Technol. Lett. 49(12), 3188–3192.
44. Conyers, J. (2010) Microstructural Kinetics Group, University of Cambridge. Microstructural Kinetics Group. [Online] http://www.msm.cam.ac.uk/mkg/e_mig.html.
45. Ruan, J. et al. (2008) ESD effects in capacitive RF MEMS switches. 8th World Congress on Computational Mechanics (WCCM8), Venice, Italy.
46. Shea, H.R. et al. (2000) Anodic oxidation and reliability of MEMS Poly-silicon electrodes at high relative humidity and high voltages. Proc. of SPIE, Vol. 4180, pp. 117–120.
47. Wada, Y., Nishimatsu, S. (1987) Grain Growth Mechanism of heavily phosphorus-implanted polycristalline silicon. J. Electrochem. Soc. 125(9), 1499–1504.
48. Kahn, H. et al. (2002) Fatigue failure in polysilicon not due to simple stress corrosion cracking. Science. 298, 8.
49. Fitzgerald, A.M. et al. (2002) Subcritical crack growth in single-crystal silicon using micromachined specimens. J. Mater. Res. 17(3), 683–692.

50. Shea, H.R. et al. (2004) Effects of electrical leakage currents on MEMS reliability and performance. IEEE Trans. Device Mater. Reliab. 4(2), 198–206.
51. The International Technology Roadmap for Semiconductors. (2000).
52. Walraven, J.A. et al. (2008) *The Sandia MEMS Passive Shock Sensor: FY08 Failure Analysis Activities.* Albuquerque, NM: Sandia National Labs, SAND2008-5185.
53. Mac Gibbon, B.S., Busnaina, A., Fardi, B. (1999) The effect of thermophoresis on particle deposition in a low pressure CVD reactor. J. Electrochemical Soc. Solid State Sci. Technol. 146.
54. Bhushan, B. (2004) *Springer Handbook of Nanotechnology.* Heidelberg: Springer.
55. Bhushan, B. (1999) *Principles and Applications of Tribology.* New York: Wiley.
56. Derjaguin, B.V., Muller, V.M., Toprov, Y.P. (1975) Effect of contact deformation on the adhesion of particles. J. Colloid Interface Sci. 53(2), 314–326.
57. Li, H. et al. (1999) Using TOF-SIMS to inspect copper-patterned wafers for metal contamination. MicroMagazine.com. http://www.micromagazine.com/archive/99/03/li.html.
58. Hyun, S., Brown, W.L., Vinci, R.P. (2003) Thickness and temperature dependence of stress relaxation in nanoscale aluminum films. Appl. Phys. Lett. 83(21), 4411–4413.
59. Maier-Schneider, D. et al. (1995) Variations in Young's modulus and intrinsic stress of LPCVD-polysilicon due to high-temperature annealing. J. Micromech. Microeng. 5(2), 121.
60. Lee, H.J., Cornella, G., Bravman, J.C. (2000) Stress relaxation of free-standing aluminum beams for microelectromechanical systems applications. Appl. Phys. Lett. 76(23), 3415–3418. ISSN: 0003-6951.
61. Maboudian, R., Howe, R.T. (1997) Critical Review: Adhesion in Surface Micromechanical Structures. J. Vac. Sci. Technol. B. 15, 1–20.
62. Mastrangelo, C.H. (1998) Surface forces induced failures in microelectromechanical systems. In B. Bhushan (ed) *Tribology Issues and Opportunities in MEMS.* Dordrecht: Kluwer Academic, pp. 367–395.
63. Bhushan, B. (1996) Tribology and Mechanics of Magnetic Storage Devices, 2nd edn. New York: Springer.
64. Mastrangelo, C.H., Hsu, C.H. (1993) Mechanical stability and adhesion of microstructures under capillary forces: Part II. Basic theory. J. Microelectromech. Syst. 2(1), 33–43.
65. Mastrangelo, C.H., Hsu, C.H. (1993) Mechanical stability and adhesion of microstructures under capillary forces: Part II. Experiments. J. Microelectromech. Syst. 2(1), 44–55.
66. Maboudian, R., Howe, R.T. (1997) Stiction reduction processes for surface micromachines. Tribol. Lett. 3, 215–221.
67. Merlijn van Spengen, V., Puers, R., De Wolf, I. (2002) A physical model to predict stiction in MEMS. J. Micromech. Microeng. 12, 702–713.
68. Boer, M.P., Michalske, T.A. (1999) Accurate method for determining adhesion of cantilever beams. J. Appl. Phys. 86(2), 817–827.
69. Knapp, J.A., Boer, M.P. (2002) Mechanics of micro-cantilever beams subject to combined electrostatic and adhesive forces. J. Microelectromech. Syst. 11(6), 754–764.
70. Mercado, L.L. et al. (2004) Mechanics-based solutions to RF MEMS switch stiction problem. IEEE Trans. Compon. Packag. Technol. 27(3), 560–568.
71. Mastrangelo, C.H. (1999) Suppression of stiction in MEMS. MRS, Boston.
72. Sparks, D. et al. (1999) Chip-scale packaging of a gyroscope using wafer bonding. Sens. Mater. 11(4), 197–207.
73. Lieberenz, T., Martin, D. (2006) Dicing before grinding for wafer thinning. Chip Scale Review. http://www.chipscalereview.com/issues/0506/article.php?type=feature&article=f3.
74. Harper, C.A. (2000) *Electronic Packaging and Interconnection Handbook.* New York: McGraw-Hill, 0-07-134745-3.

Chapter 4
In-Use Failures

4.1 Introduction

This chapter addresses in-use failures of MEMS, with an emphasis on the physics of failure. Chapter 3 dealt with eliminating failures from a design and manufacturing perspective. In this chapter we focus on how a well-designed, fabricated and packaged device can fail in use. There is a tight link between the design, manufacturing and in-use failures. Understanding the physics of failure (e.g., creep, fatigue) and the properties of materials used and the link to the process flow (e.g., yield strength of poly-silicon following HF release) lead to improved design rules to ensure the device will operate reliably in the expected operating environment. A concurrent design of the package is often required, but is not addressed in this chapter.

The chapter is organized into sections dealing with failures of mechanical origin (shock, fatigue, creep), electrical origin (dielectric charging, ESD), and related to environmental effects (radiation, anodic oxidation). For each case, the underlying physics are summarized, followed by several examples, mostly from commercial MEMS, showing the trade-offs required to obtain high reliability.

Different devices will have different predominant failure modes. As the device matures, this mode will change as successive changes in materials, geometry, process flow, packaging, actuation or sensing waveforms, increase the device reliability. Efficiently designing a MEMS device to be reliable while ensuring rapid time to market requires concurrent engineering practices, as presented in [1] by S. Arney. Understanding the physics of failure, which allows accelerated testing to be performed with high confidence, is one key element of such an approach, illustrated in Fig. 4.1.

4.2 Mechanical Failure Modes

In this section, fracture, creep (plastic deformation) and fatigue are discussed as failure modes. We do not address wear here, as it affects very few MEMS devices, since almost all designs go to great lengths to avoid rubbing MEMS parts. Due to the large surface to volume ratio for MEMS devices, and the dominance of surface forces

A.L. Hartzell et al., *MEMS Reliability*, MEMS Reference Shelf,
DOI 10.1007/978-1-4419-6018-4_4, © Springer Science+Business Media, LLC 2011

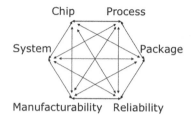

Fig. 4.1 Design for reliability approach, illustrating the interdependence of MEMS product development activities. Reference [1] reproduced by permission of the MRS Bulletin, Copyright 2001 Materials Research Society

over inertial forces at the μm size scale, wear can be a particularly acute problem for MEMS. It has been extensively studied by Sandia National Laboratories, and comprehensively documented in [2]. We do not covered here as it can be avoided for nearly all types of MEMS by careful design to avoid rubbing parts. Stiction is a related but more widespread failure mode, and has been covered in Chapter 3.

4.2.1 Fracture

Fracture can be an important failure mode for MEMS devices, which, by definition, include micromachined mechanical components. There is the widespread perception outside the MEMS field that silicon MEMS are fragile devices, because silicon is a brittle material. Yet Silicon is a beautiful mechanical material, as long as one designs the device to operate well below the facture strength, as illustrated in Fig. 4.2 for SOI micromirrors from Alcatel-Lucent USA Inc., which were externally deformed in ways that would never happen in normal use to show the elastic behavior of well-engineered silicon suspensions designed with large safety factors.

An important element distinguishing the mechanical design of MEMS from the mechanical design of macroscopic sensors and actuators is that the properties of thin films can differ significantly from that of bulk samples, because the film thickness is comparable to grain size, or because of process-related damage or modifications.

For this reason, there has been a large body of work to measure the mechanical properties of the thin films used in MEMS: polysilicon, SiN, metals, etc. The properties measured, e.g., Young's modulus, fracture strength, yield strength, fracture toughness, Poisson ratio, have been reported for all main elements and films used in MEMS, for instance in Chapter 3 of the MEMS Handbook (CRC press, 2002) [3] and the MEMSNET website [4].

Figures 4.3 and 4.4 illustrate schematically the stress-strain curves of brittle and ductile material. In both cases the slope of the linear (elastic) regime is the Young's modulus. But for brittle materials, behavior is elastic up till fracture, while ductile materials exhibit a larger region of plastic deformation.

We shall limit our brief discussion to silicon, whose fracture strength, as for all brittle materials, is distributed following Weibull statistics (see Chapter 2), as fracture initiation relies on a pre-existing defect or crack. Figure 4.5 shows the influence of surface area of a polysilicon beam on Tensile strength. There are

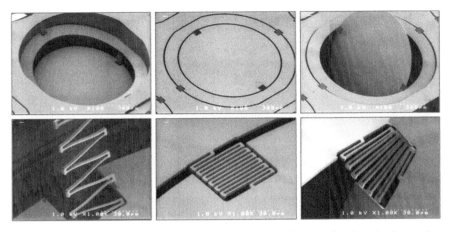

Fig. 4.2 Single crystal silicon micro-mirrors from Alcatel-Lucent showing the impressive mechanical properties of silicon. *Center*: mirror at rest position. The *right* and *left* images were taken after using a probe needle to move the mirrors to positions where the stress in the springs is one order of magnitude larger than what would be encountered in normal (electrostatic actuation) operation, yet because of careful engineering and large safety margins, the springs deform elastically and do not fracture. Reprinted with permission of Alcatel-Lucent USA Inc.

Fig. 4.3 Schematic stress-strain curve for a brittle material. The elastic limit corresponds to the ultimate strength, beyond which the device fractures

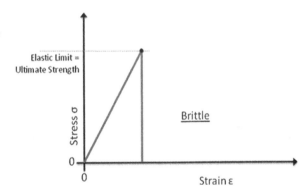

Fig. 4.4 Schematic stress-strain curve for a ductile material. The elastic limit corresponds to the yield strength, after which the material deforms plastically. The ultimate strength is the maximum of the stress-strain curve, and fracture occurs at the breaking strength

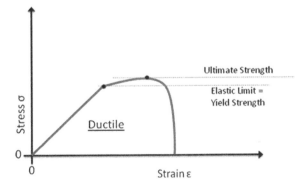

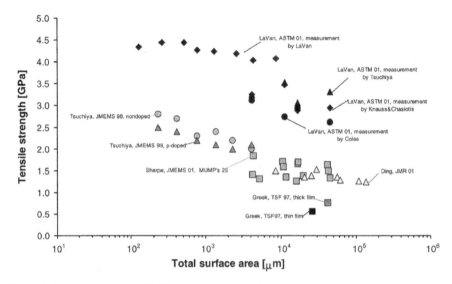

Fig. 4.5 Tensile strength of polysilicon vs. beam area, showing large differences between different fabrication techniques, and also showing higher fracture strengths when smaller or areas are involved, indicating that surface defects rather than volume defects are the initiating points for fracture. Reference [8] reprinted with permission Copyright 2003 IEEE

two important observations: (1) smaller area beam have a higher tensile strength (because there is a smaller area for pre-exiting cracks or damage), and (2) the properties of polysilicon differ widely depending on growth conditions, doping, grain size, surface roughness, etc.

To ensure sufficient design margin, it is essential to know the properties of the films one is using, though this is easier for single crystal silicon. Figure 4.6 from [5] plots the Weibull failure probability for each of the five layer of Sandia's SUMMiT V^{TM} process, showing both that within a layer there can be nearly a factor of two range in failure stresses, and that the average failure stress varies by a factor of 2. It is important to use a large safety margin, and obtain test data on the specific polysilicon layer one is using.

Processing play a large role in fracture of brittle materials, as it leads to the formation of initial cracks (for instance, see the example of pitting of polysilicon in Fig. 5.17, leading to stress concentration). It depends also on crystal orientation. The data in Table 4.1 from Chen et al. [6] is for single crystal silicon, showing Weibull reference strength ranging from 1.2 GPa for mechanically ground <100> silicon wafers to 4.6 GPa for the same silicon once etched by DRIE. This variation must be taken into account when designing a MEMS device, to ensure the correct yield strength is used. Simply looking up the generic value for a given material is not sufficient if high reliability is sought after. Very conservative design margins, as was the case for the Alcatel-Lucent micromirrors shown in Fig. 4.2, is another route to ensure devices will not fracture.

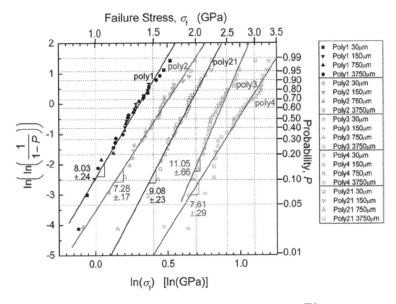

Fig. 4.6 Weibull failure probability plot for each of five SUMMiT V™ layers. Reference [5] reprinted with permission Copyright 2007 IEEE

Table 4.1 Table of strength characteristics of single crystal <100> silicon with different surface conditions

	Mechanically ground (A)	Mechanically ground (B)	KOH-etched silicon (C)	STS DRIE ailicon (D)	Chemically polished
Sample size	19	30	25	20	10
Specimen thickness (μm)	500	280	280	230	280
P-P surface roughness (μm)	∼3	∼1	∼0.3	∼0.3	∼0.1
σ_0 (GPa)	1.2	2.3	3.4	4.6	>4
Weibull modulus m	2.7	3.4–4.2	7.2–12	3.3	?
Effective R_v for uniaxial volume specimen (mm)	0.383	0.284–0.215	0.12–0.102	0.295	
Effective R_A for uniaxial surface specimen (mm)	1.94	1.266–0.857	0.421–0.487	1.339	

[†]Note: The volume of the equivalent uniaxial volumetric flaw specimen is $\pi\ R^2_v h$, where $h =$ 300 μm. The surface area of the equivalent uniaxial surface flaw specimen is $2\pi R_A h$.
Adapted from [6] reprinted with permission Copyright 2000 John Wiley and Sons.

Brittle materials exhibit a transition to ductile behavior as the temperature is increased and dislocation motion becomes possible. The data generally reported for MEMS materials is measured at room temperature. For harsh conditions (e.g., a pressure sensor near combustion chamber) where high temperatures are expected, there will be a marked reduction in yield strength, see Fig. 4.7 for polysilicon. Sharpe et al. [7] have shown that Si can exhibit large ductility when heated to over 700°C, as shown the in the stress-strain curve in Fig. 4.8.

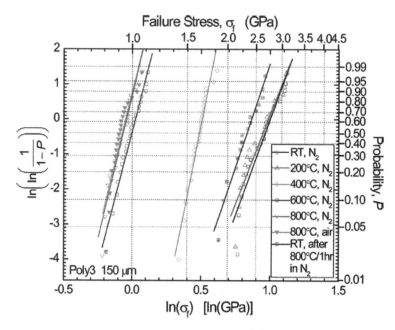

Fig. 4.7 Weibull failure probability plot for one SUMMiT V™ polysilicon layer for several temperatures up to 800°C, showing a strong reduction in failure stress at elevated temperature. Adapted from [5] reprinted with permission Copyright 2007 IEEE

4.2.2 Mechanical Shock Resistance

4.2.2.1 Introduction

Shock is a sudden acceleration. Rather than using SI units of ms^{-2}, shocks are commonly described by the peak acceleration, in units of "g", where $1\ g = 9.81\ ms^{-2}$, the acceleration in Earth's gravity.

While the normal operating environment for most consumer devices is 1 g, these devices actually need to withstand large shocks in order to have an acceptable lifetime, i.e., in order to be reliable. Devices can be accidentally shocked, for instance being dropped (e.g., a mobile phone falling out of a pocket, a component

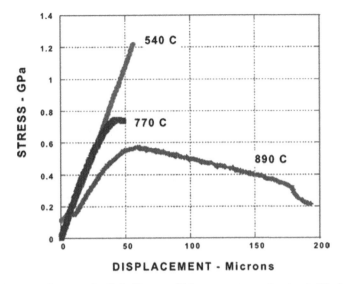

Fig. 4.8 Stress-strain curve for Polysilicon at high temperatures, showing brittle behavior at 540°C, but ductile behavior at 890°C. Reference [7] reprinted with permission Copyright 2003 Springer

falling on the floor during assembly, roughly 500 g for a 1.5 m fall on a hard surface), handled roughly (a mechanical wristwatch worn while playing tennis). Some devices are designed to operate in high shock environments, such as accelerometers in cars which might need to sense shocks of up to 100 g. More extreme examples involve devices on spacecraft, which may need to survive the shock of pyrotechnic bolts used for stage separation (>10,000 g) or of landing (of order 30 g for airbag landing on Mars), and for arming and safeing devices for ammunition fired from a gun or artillery, where shocks of up to 100,000 g are reported.

The failure modes induced by shock include:

- Fracture due to exceeding the yield strength of the material because of a large shock-induced deflection. This is the most obvious failure mode, and will be the main one discussed in the section
- Stiction due to parts coming into contact that would not do so under normal operation. This is termed in-use stiction.
- Delamination, e.g., due to die attach failure, or between layers in a surface micromachined structure
- Particulates being generated or being displaced, leading both to the short circuits between neighboring electrodes, as well as mechanical blockage, e.g., micron-size particles blocking or short-circuiting a comb drive.
- Short-circuits due to parts at different potentials (e.g., comb fingers) coming into contact because of mechanical shock.

From a scaling perspective, MEMS are much more shock tolerant than larger devices for failures due to fracture. A dimensional analysis shows the critical shock scales as 1/L, where L is a typical dimension (e.g., length). The force on a proof mass due to an acceleration a is simply $F=ma$, and for micron-scale devices, the mass is extremely small. We shall return to this in more detail in Section 4.2.2.4. The situation is slightly more complicated than simply stating the masses are small: in order for an inertial sensor to have a useful sensitivity, it must move by a least a few nm under the applied acceleration, which implies a sufficiently compliant suspension, hence possibly susceptibility to mechanical shock. Similarly MEMS actuators (such as micromirrors) cannot have overly rigid suspension, as this would require an overly large driving force, which would lead to very high drive voltages in the case of electrostatic actuation. Selecting a compliant suspension to increase sensitivity can lead to higher shock susceptibility.

The package of course plays a central role in transmitting shock from the environment to the device. The type of die attach (solder, glue) and the way the package itself is mounted to the shock tester provide different transfer functions generally attenuating the shock but possibly amplifying it at a resonance, and also changing the shock duration. The package and die-attach can fail too (often more likely to fail than MEMS because of larger mass), as discussed further below.

It is important from a product perspective to distinguish between *shock survival* which is the maximum shock that a device can support before complete failure, and *shock resistance* which is the maximum shock it can support without degrading the specifications beyond what is described in the data sheet (for instance the shift of the scale factor or shift of the bias). Shock resistance will correspond to a lower acceleration than shock survival, for instance because shocks could lead to a shift in the die attach and hence to stress on a sensor chip, changing the calibration.

Since MEMS can withstand, when suitably designed and packaged, very large shocks, testing them can be a challenge. Shock levels of up to 6000 g are readily achieved with standard drop-test shock testers and hammers, but larger shocks require specialized equipment, for instance Hopkinson bars (e.g., [9]) and ballistic tests are needed to reach the 10,000–100,000 g levels. One might test devices at shocks higher than the maximum expected operating level in order to have sufficient safety margin and in order to find out at what at shock level the devices fail, and how they fail.

The very high shock levels tend only to be of concern for military applications dealing with smart munitions guidance (inertial sensors) or arming and safeing. For the later, a safe and arm mechanism is a means to ensure ordnance only explode after being fired: to prevent warhead explosion prior to firing, the munitions are only armed once a longitudinal acceleration of, for instance, 20,000 g is detected, and a radial acceleration due to the spin of the projectile of 10,000 g is sensed. Axun Technologies, Billerica, MA, USA, has developed LIGA MEMS structures that, when assembled, only latch into position following such large accelerations [10].

4.2.2.2 Response to Shocks

Shock Modeled Using the Half-Sine Wave Approximation

Shocks are described by acceleration versus time curve, which for real-world shocks will be a complex shape. As described in [11], shock pulses can be approximated by a series of half-sine pulses. The peak acceleration a_{peak} is often referred to as the shock level, and the duration τ of the half-sine is taken to be the duration of the shock, see Fig. 4.9. Figure 4.10 plots measured acceleration versus time for two shock testing equipment at Sandia national laboratories: a shock table at 500 g, and a Hopkinson bar at 22,000 g.

The higher the peak acceleration, the shorter its duration. Several testing and qualification standards are discussed in Chapter 6 for shock and vibration. Table 4.2 shows the test conditions for mechanical shock testing according to MIL standard 883, showing a clear reduction in pulse duration with increased g level. The same relation can also be seen when comparing Fig. 4.10a, b. Reported durations range from 50 μs to 6 ms [11].

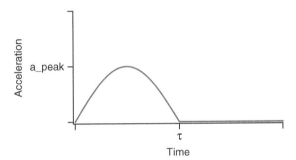

Fig. 4.9 Half-sine acceleration pulse, the acceleration is zero at time zero, reaches a maximum at time $\tau/2$, and is zero for times larger than τ

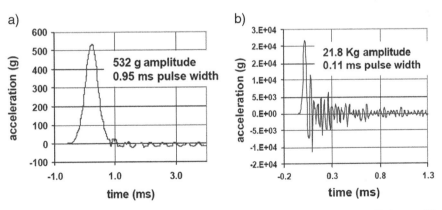

Fig. 4.10 Measured shocks on: (**a**) a shock table, and (**b**) using a Hopkinson bar. Reference [9] reprinted with permission Copyright 2000 IEEE

Table 4.2 Test conditions for mechanical shock testing per Mil-Std-883 Method 2002

Test condition	g Level (peak)	Pulse duration (ms)
A	500	1
B	1500	0.5
C	3000	0.3
D	5000	0.3
E	10,000	0.2
F	20,000	0.2
G	30,000	0.12

Srikar and Senturia [11] showed that shock response is governed by the relative magnitude of three time frames: (1) the acoustic transit time $t_{acoustic}$, (2) the time of a vibration T_{vib} (generally much larger than $t_{acoustic}$), and (3) the rise time of the applied shock τ. For $\tau < t_{acoustic}$, the system response is described by superposition of traveling elastic waves. When $\tau \sim T_{vib}$, the system response is best described using normal mode solutions. When $\tau \gg T_{vib}$, the response is quasistatic, as illustrated in Fig. 4.11.

Fig. 4.11 Relevant times scales for shock loaded MEMS, from [11] reprinted with permission Copyright 2002 IEEE

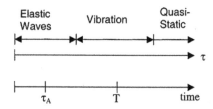

$t_{acoustic}$ is the time taken for an elastic waves to propagate through the microstructure. If we consider a simplified microstructure of length L the speed of sound is c we have $t_{acoustic} = L/c$. The fundamental resonant frequency $w_0 = 2\pi / T_{vib}$. For typical silicon MEMS, $t_{acoustic}$ is typically less than 0.1 μs, T_{vib} ranges from 0.1 μs to 0.1 s, and τ from 40 μs to 6 ms. In view of these timescales, elastic wave propagation does not play a role in the reliability of MEMS in response to mechanical shocks.

Srikar and Senturia [11] define three possible cases depending on the relative magnitudes of T_{vib} and τ.

- $\tau < 0.25\,T_{vib}$: *Impulse response.* The MEMS device responds as if it had an initial velocity equal to the integral of the acceleration pulse, i.e., the momentum of the mass is equal to the impulse of the force. The frequency of the device dictates the dynamics, and the exact value of τ does not play a role.
- $0.25\,T_{vib} < \tau < 2.5\,T_{vib}$: *Resonant response.* The shock can excite a resonance of the device, and the acceleration of the microsystem can exceed the peak applied acceleration, depending on the quality factor of the mode.

- $\tau > 2.5\ T_{\mathrm{vib}}$: *Quasistatic response*. The frequency of shock force dictates the dynamics, as the device simply tracks the applied load, given that is can respond much faster than the load varies.

Figure 4.12 illustrates these three different cases for a number of accelerometers, showing that most of the devices are in the quasistatic regime. Understanding the device dynamics is essential for understanding failure modes: if one can predict the maximum displacement for shock along any axis, one can determine whether fracture, stiction, or delamination will be an issue. Particulate motion is more difficult to compute, but avoiding impacts due to shock also avoids particular generation from those impacts.

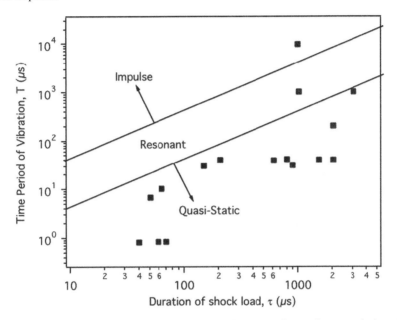

Fig. 4.12 Mechanical response of shock-loaded MEMS (principally accelerometers), from [11] reprinted with permission Copyright 2002 IEEE

For the simple (but instructive) case of a mass-spring-damper system attached to a support, the response of the mass to a half-sine shock pulse to the support can be solved analytically [12]. Consider a mass *mass*, a spring constant k, damping coefficient c_{damper}, with natural resonant frequency $\omega_{\mathrm{res}} = \sqrt{k/mass}$ (see Fig. 4.32 in Section 4.2.3). The Damping constant is $\zeta = \sqrt{c_{\mathrm{damper}}/4m\omega_{\mathrm{res}}}$. Let the half-sine pulse acceleration of the support be given by:

$$\ddot{y}(t) = \begin{cases} a_{\mathrm{peak}} \sin \frac{\pi t}{\tau}; & 0 < t < \tau \\ 0; & t > \tau \end{cases}$$

where a_{peak} is the peak acceleration and τ is the duration of the half-sine pulse. y is the displacement of the support. We assume the mass starts from rest, and solve for the relative motion of the mass $z(t)=x(t)-y(t)$, which we write as:

$$z(t) = \begin{cases} R(t) \; ; & 0 < t < \tau \\ R(t) + R(t - \tau) \; ; & t > \tau \end{cases}$$

For compactness of the final equation for $R(t)$, we define the following intermediate variables:

$$a = \pi^2/\tau^2, \; b = 2\zeta\omega_{res}, \; c = \omega_{res}^2, \; \omega_d = \omega_{res}\sqrt{1 - \zeta^2} \; \text{(only for } \zeta < 1)$$

$$l = \frac{-b}{(a - c)^2 + ab^2}; \; m = \frac{c - a}{(a - c)^2 + ab^2}; \; n = \frac{b}{(a - c)^2 + ab^2}; \; p = \frac{b^2 - c + a}{(a - c)^2 + ab^2}$$

$u(t)$ is the unit step (or Heaviside) function.

The exact solution depends on the value of the damping constant and is hence split into three cases: under-damped ($\zeta<1$), critically damped ($\zeta=1$) and over-damped ($\zeta>1$). One obtains:

1. For an *underdamped* system ($\zeta<1$), $R(t)$ can be written as:

$$R(t)_{underdamped} = -a_{peak}\frac{\pi}{\tau}\left(l\cos\left(\frac{\pi}{\tau}t\right) + \frac{m\tau}{\pi}\sin\left(\frac{\pi}{\tau}t\right)\right)$$
$$+e^{-\zeta\omega_{res}t}\left(n\cos(\omega_d t) + \frac{p-n\zeta\omega_{res}}{\omega_d}\sin(\omega_d t)\right)\right) \cdot u(t) \quad (4.1)$$

2. For a *critically-damped* system ($\zeta=1$), $R(t)$ can be written as:

$$R(t)_{crit.damped} = -a_{peak}\frac{\pi}{\tau}\left(l\cos\left(\frac{\pi}{\tau}t\right) + \frac{m\tau}{\pi}\sin\left(\frac{\pi}{\tau}t\right)\right.$$
$$\left. +e^{-\omega_{res}t}\left(n + (p - n\omega_{res})t\right)\right) \cdot u(t) \quad (4.2)$$

3. For an *overdamped* system ($\zeta>1$), $R(t)$ can be written as:

$$R(t)_{overdamped} = -a_{peak}\frac{\pi}{\tau}\left(l\cos\left(\frac{\pi}{\tau}t\right) + \frac{m\tau}{\pi}\sin\left(\frac{\pi}{\tau}t\right) + Je^{-\left(\zeta-\sqrt{\zeta^2-1}\right)\omega_n t}\right.$$
$$\left. +Ke^{-\left(\zeta+\sqrt{\zeta^2-1}\right)\omega_n t}\right) \cdot u(t)$$
$$(4.3)$$

$$\text{where } J = -\frac{n\omega_{res}\left(\zeta - \sqrt{\zeta^2 - 1}\right) - p}{2\omega_{res}\sqrt{\zeta^2 - 1}} \text{ and } K = \frac{n\omega_{res}\left(\zeta + \sqrt{\zeta^2 - 1}\right) - p}{2\omega_{res}\sqrt{\zeta^2 - 1}}$$

The above equations allow the maximum mass displacement to be computed vs. shock level and pulse duration for simple systems, and hence provide a first tool to judge whether for instance stiction is possible (do two parts collide?) and determine the kinetic energy of MEMS parts at collision. Figure 4.13 illustrate the displacements for different damping conditions using (4.1) to (4.3).

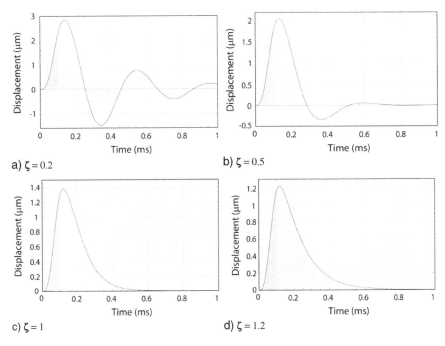

Fig. 4.13 Displacement of mass relative to substrate for an 100 μs long, 100 g peak amplitude shock pulse, for a device with an undamped resonant frequency of 2.5 kHz, for damping coefficient of (**a**) 0.2, (**b**) 0.5, (**c**) 1, (**d**) 1.2. Displacement was determined using equations (4.1) (**a**, **b**), (4.2) (**c**) and (4.3) (**d**). The change in shading indicates the end of the shock pulse at $t=100$ μs. Note the larger maximum displacement for devices with less damping

High-g Shock Data on Silicon MEMS

Sandia National Laboratories have published extensively on the shock testing of their micro engine devices, which are complex and rather large surface micromachined polysilicon devices incorporating comb drives, gears, linkages, masses, and springs. In [9], Tanner et al. report a variety of failure modes in response to shocks ranging from 500 to 40,000 g, applying the shocks on all three axes. At 1000 g, no damage was observed. At 4000 g debris (small particulates) on the surface of the die were observed to move slightly. At 10,000 g, 90% of the 19 devices were still operational. The observed failures were due to delamination: the die attach failed, because it had been weakened by the coupling agent used to prevent stiction in the MEMS release step. Only at 20,000 g did the MEMS devices begin to fail due to fracture of polysilicon components. Debris also move at the shock levels and can lead to short-circuits as shown in Fig. 4.14. At 40,000 g the packages failed, see Fig. 4.15. Amazingly several die that were removed from the fractured packages were operational when placed under a probe station. This data shows that even large MEMS devices can be very shock tolerant. This report also indicates that the direction of the shock is a key parameter.

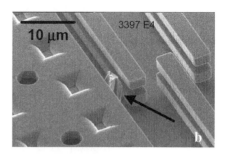

Fig. 4.14 Particulate
contamination following
20,000 g shock on a Sandia
National laboratories micro
engine [9] reprinted with
permission Copyright 2000
IEEE

Fig. 4.15 ceramic package following 40,000 g impact. The die attach is visible at the center of the recess. The die was removed and found to still be functioning despite the large amount of debris [9]. Reprinted with permission Copyright 2000 IEEE

The fracture of brittle materials under tension or bending depends on defects initiating the crack, and thus generally follow a Weibull distribution, as reported for instance in [13] and shown in Fig. 4.6. The same statistics should therefore hold for fracture due to applied shocks. Indeed Wagner et al. [14] have investigated the response of epi-polysilicon to mechanical shocks, and reported an excellent fit to Weibull statistics of the measured cumulative failure probability versus peak acceleration, see Fig. 4.16. This data shows the wide range of forces that lead to

Fig. 4.16 Weibull plot of
cumulative failure frequency
vs. peak acceleration, for
epi-polysilicon MEMS device
consisting of a 10 μm thick
1 mm^2 proof mass suspended
by beam of width less than
5.6 μm subjected to tensile
load by repeated shocks [14].
Reprinted with permission
Copyright 2003 IEEE

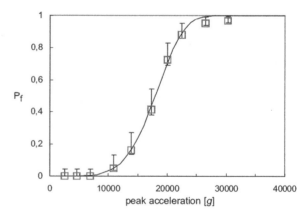

fracture for brittle materials, and emphasize the need for suitably large number of samples for testing, or suitably large safety margins to account for the range in fracture strengths of a given brittle material.

4.2.2.3 Increasing Shock Resistance

Recalling the main failures modes related to mechanical shock (fracture, stiction, delamination, particulates, short-circuits), what can be done to increase the shock tolerance of the device?

A commonly implemented solution that addresses fracture and delamination is the use of stoppers to limit the deflection of the moving parts. Figure 4.17 illustrates the use of a stopper to provide 2-D displacement limits for an Analog Devices accelerometer. A stopper is generally made from the same material as the MEMS mass, and is easy to implement, as it requires simply a mask change.

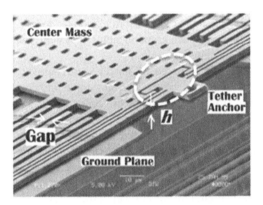

Fig. 4.17 (*Left*) SEM image of one suspension of the Analog Devices ADXL76 accelerometer. Capacitive sense comb-finger are visible on both sides of the center mass. The stopper is the T-*shaped* structure in the dashed circle, and limits the in-plane motion of the center mass to avoid contact between fingers of the comb drives. Reference [17]. Reprinted with permission. Copyright 1999 IEEE

By limiting the motion, the stopper avoids fracture by ensuring the stress in the device is below the fracture strength (or the elastic limit). It similarly reduces delamination. The range of motion allowed by the stopper determines the maximum kinetic energy E_c of the mass which was shown in reference [9] to be simply $E_c = mad$, where m is the mass of the moving object, a is the acceleration, and d the gap at rest between stopper and moving mass. To avoid chipping and particulates generation, one should minimize E_c, and hence select as small a gap as is possible given the fabrication technology and required displacement of the device under normal operation. Figure 4.18 is an SEM image of a silicon stopper, showing debris due to impact. The debris are particularly worrisome in the shock environments in view of their mobility.

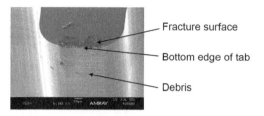

Fig. 4.18 Chipping on silicon stopper following impact, showing fracture surface and accumulated debris [18]. Reprinted with permission Copyright 2008 Society of Photo Optical Instrumentation Engineers

Two limitations of the use of stoppers are stiction and particulate generation. Since hard shock stops can lead to impact, debris and oscillations there has been some modeling of softer stoppers, either using non-linear springs to minimize impact, or adding coatings with low coefficient of restitution such as polymers or metals including gold and copper [15]. More elaborate stoppers based on the techniques used in watchmaking, such as the Incabloc® system, can be adapted MEMS and provide much greater safety margin.

Anti-stiction coatings as discussed in Chapter 3, are routinely employed to ensure reliable operation of devices where services may come into contact, such as surface micro-machined accelerometers, of which an Analog Devices model was presented as a case study in Chapter 2. For the Analog Devices accelerometers, the final step before the sensor element is released is the LPCVD (vapor phase) deposition of an organic anti-stiction coating approximately 0.8 nm thick (the process is self-limiting). The coating greatly reduces surface energy, and hence increase the likelihood that parts will not stick should they come into contact (which they should not under normal circumstances), yet is thin enough that it does not affect the mechanics of the devices. Texas instruments also use vapor-phase deposited anti-stiction coatings in its DMD chips, self-assembled monolayers of $CF_3(CF_2)_8COOH$ have been reported [16]. Several anti-stiction coatings are discussed in Chapter 5, illustrating analytical methods for failure analysis.

We now discuss briefly by way of example an accelerometer designed for operation at high shocks (20,000 g according the to the data sheet [19]), the Colibrys SA (Neuchâtel, Switzerland) model HS8030, similar to the MS 8000 shown in Fig. 4.19. Colibrys had developed several generations of capacitive accelerometers, packaged in ceramic multichip modules, with full-scale ranges from 2 to 200 g. In order to ensure high performance the MEMS sensor chip must be attached to the ceramic carrier with a compliant die-attach so as not to apply stress to the sensing chip from mounting the package, or internally due to CTE mismatch between silicon chip and ceramic package as the device needs to operate without bias change over a temperature range of −55°C to +125°C.

Compared to the conventional product, with stoppers designed to limit motion of the proof mass with respect to the MEMS chip, it was necessary, because of the compliant die attach, to also implement stoppers limiting the motion of the MEMS

Fig. 4.19 Colibrys HS 8000 series accelerometer. (*Left*) packaged, (*right*) MS 8000 without package lid, showing the sensor chip and analog and digital signal conditioning circuits. Courtesy Colibrys SA

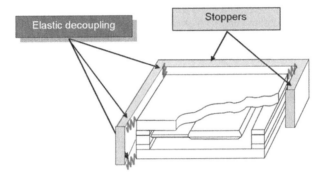

Fig. 4.20 Schematic isometric diagram of the Colibrys HS8000 accelerometer. The proof mass moves vertically in response to acceleration. Stoppers are implemented to limit motion of the chip. Courtesy Colibrys SA (patent pending)

chip itself [20], see Fig. 4.20. The main challenge was not the shock resistance of the sensor, which was shown to survive 40,000 g, but the need to combine Colibrys' existing patented stress isolation technique (required to meet the demanding performance specifications) with stoppers to form a hybrid shock protection solution [21], see Fig. 4.21.

Standard testing equipments are generally not capable to reproduce extreme conditions. Certain harsh conditions can be replicated by a combination of standard tests, for instance combining a hammer test with a centrifuge to simulate a gun hard shock. To test the accelerometers at high g levels (gun hard, 20,000 g, 10 ms), Colibrys used a shock test equipment known as an aerobutt tester at BAE Systems, depicted in Fig. 4.22, which provides much longer duration pulses than the Hopkinson bar data in Fig. 4.10b.

Colibrys tested 124 accelerometers (model HS8030, 30 g full scale) in the Aerobutt tester. The resulting shift in bias had a 3 σ variation of under 50 mg, thus demonstrating the effectiveness of the hybrid mounting solution to increase shock resistance of the packaged device to 20,000 g.

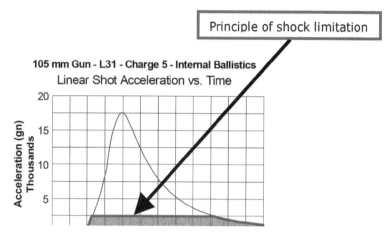

Fig. 4.21 Principle of a shock limitation by a combination of the elastic decoupling and stoppers, to limit the shock at the chip level to 2500 g while the package is accelerated to over 17,000 g. Courtesy Colibrys SA

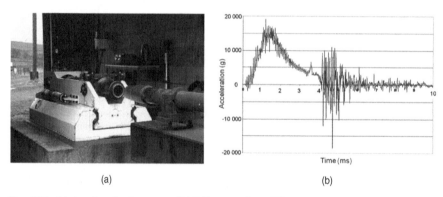

(a)   (b)

Fig. 4.22 (**a**) Aerobut shock tester at BAE Systems, from [22] Reprinted with permission copyright 2006 IEEE, (**b**) measured shock pulse, showing the complexity of real testing compared to the simple half-sine description, Courtesy Colibrys SA

Reliability of Low-G Sensors

There is increasing interest in extremely sensitive inertial sensors capable of operating in the nano-g regime (below 10^{-8} ms^{-2}), for instance for seismic sensing as well as for some in-orbit applications. Because the suspension of such a device must be extremely compliant in order to get at least a few nanometers deflection due to the desired stimulus, the critical acceleration (at which fracture occurs) for these devices is only of order 10 g. So merely handling such a device during fabrication can lead to failures, since picking up an object and putting it down easily generates accelerations of over 20 g.

Such devices generally have springs that are highly compliant only in one direction, and much stiffer in the other two. These devices need suitable stoppers to prevent excessive motion, but the challenge is integrating stoppers on three axes prior to packaging but after release. The solution implemented first at Sandia National Laboratories and then at the EPFL were to package the chip prior to final release, so that the proof mass is either locked in place by silicon oxide, or constrained by stoppers, and hence can be handled safely at all steps of the fabrication and after final release (here release means etching of the sacrificial silicon oxide holding the proof mass in place).

Figure 4.23 shows a schematic cross-section of the final assembly and HF vapor etch, and an optical micrograph of a completed inertial sensor to measure the gravity gradient vector [23]. The wafer is diced by a vapor HF step following font and backside DRIE. The chip is then bonded to a Pyrex chip to provide a bottom stopper. The top stopper is fabricated from overhangs in the top device layer. The 4 cm long MEMS proof mass is only released (another HF vapor etch) after the bottom Pyrex chip is bonded. Despite have a 1 mm long and only 5 μm wide beam suspending a 0.35 g silicon proof mass, the device can safely be handled with no special care thanks to the stoppers integrated in the 0-level package.

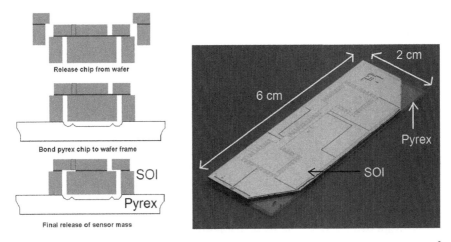

Fig. 4.23 Inertial sensor for measuring the gravity gradient in Earth orbit. The large (1×4 cm^2 footprint) proof mass is suspended by a very soft spring, and $a_{critical}$ is only 2 g. Yet by suitable constraints in 3 axes prior to final HF release the device can easily be handled. Reference [23] reprinted with permission Copyright 2009 IEEE

4.2.2.4 Simple Model for Critical Acceleration and Case Study on SOI Micro-Mirrors

In this section we will develop a simple model to estimate the maximum acceleration that MEMS devices can sustain, assuming failure occurs due to beam fracture (ignoring stiction, short circuits, delamination, etc). Using basic mechanics will calculate the strain in suspension beam as a function of acceleration. Knowing the yield strength of typical MEMS materials then allows for a quick determination of the maximum acceleration the system can take. The model developed in this section is more appropriate for static high-G loading in a centrifuge than for shock testing, as it ignores rise time and time period of vibrations, but provides an easy way to compare different geometries. Since we have seen above that many MEMS devices respond quasi-statically to shock (when $T_{vib} < 0.4 \; \tau$), the calculation does hold general interest. A similar calculation is given by Tanner et al. in ref [9].

It is recognized that the dynamic stress, and not simply the maximum acceleration as a better criterion for determining dynamic strength [24]. A correct calculation will take into account the mode of the device that is excited, and hence the coupling of energy into this mode as a function of the duration of the shock pulse, see equations (4.1), (4.2), and (4.3) above for different damping conditions. The effect of the package must also be taken into account, in particular for a device falling onto a hard surface. For simplicity we shall ignore these issues and focus on finding order of magnitude values that serve as starting points for a more complete calculation. The "Shock and Vibration Handbook", in particular Chapter 8 on transient response to step and pulse functions [25] provides a good framework for a more complete solution.

We will show that the model fits well data on some micro mirrors from Alcatel-Lucent USA Inc. but also show how the model fails for other micro mirrors when the model is overly simplistic. We shall give some examples showing how springs can be re-engineered to deal with stress concentration and illustrate how shock on all three axes must be considered to come up with a reliable design.

Cantilever Beam with Mass at the End

Suppose we have a mass m suspended at the end of a cantilever beam, as shown in Fig. 4.24, and that this simple device experiences an acceleration a. A shock is indeed not a constant acceleration, as discussed in the section above, but a static calculation is much simpler and still allows a good approximation of the maximum stress in the beams. We shall ignore the mass of the cantilever.

There are two forces on the mass: the restoring force of the cantilever, and the force due to the acceleration $F = ma$. Rather than deriving all the equations, we

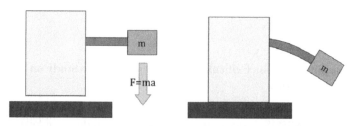

Fig. 4.24 Mass m at the end of a cantilever deforming under the influence of acceleration

will use the formulas that are very conveniently given in the helpful reference book "Roark's Formulas for Stress and Strain" [26].

The stress σ in a beam is given by $\sigma = \frac{Mz}{I}$ where M is the bending moment of the applied force, z is the distance measured from the neutral plane, and I is the moment of inertia of the section of the beam. For a beam of rectangular cross section, we have:

$$I = \frac{1}{12}t^3w$$

where t is the thickness, and w the width, and the bending is in the direction of the thickness. The maximum stress will be on the top and bottom of the beam, at $z = \pm t/2$, see Fig. 4.25.

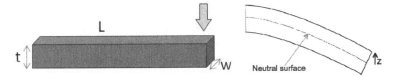

Fig. 4.25 Geometry of the cantilever, with the load applied as indicated by the *arrows*, so the neutral plane is located at $t/2$

The maximum moment for a beam of length L is:

$$M_{\max} = -FL$$
$$= -mal$$

The maximum stress is then:

$$\sigma_{\max} = \frac{M}{I}\frac{t}{2}$$
$$= -maL\frac{6}{wt^2}$$

Neglecting fatigue, creep and stiction, we expect failure when $\sigma_{\max}$ is equal to the yield strength (keeping in mind that fracture of brittle materials follows a Weibull distribution and therefore a significant safety margin is needed, see the section on fracture earlier in this chapter). One can then determine the acceleration a_{critical} when the beam fractures:

$$a_{\mathrm{critical}} = \sigma_{\mathrm{yield}}\frac{wt^2}{6\,mL} \tag{4.4}$$

From (4.4), it follows that to be able to withstand higher accelerations, the beam must be made shorter, thicker and wider. In view of the t^2 scaling, increasing the thickness of the cantilever can be the most effective technique.

For polysilicon the yield strength ranges from approximately 1 to 3 GPa, and is roughly 5 to 8 GPa for single crystal silicon [27]. As an example, 1 μg mass (e.g., a silicon cube 80 μm on a side) at the end of a 400 μm long polysilicon beam, 2 μm wide and 2 μm thick, has a maximum acceleration $a_{\text{critical}} = 10{,}000 \text{ m/s}^2 = 1000$ g.

Or, more realistic as comparable to some commercially available bulk micromachined accelerometers, a silicon proof mass, $1 \times 1 \text{ mm}^2 \times 0.5$ mm thick, suspended from 50 μm long, 50 μm wide 20 μm thick suspension, has a maximum acceleration $a_{\text{critical}} = 250{,}000 \text{ m/s}^2 = 25{,}000$ g. In practice, such a system would have stoppers to limit the motion and hence the kinetic energy, and the mass would hit those hard stops.

Note that while the deflection Δh at the end of the beam depends on the Young's modulus E as:

$$\Delta h = \frac{maL^3}{3EI}$$

the critical acceleration does not depend directly on E, but only on yield strength. a_{crit} and E are nevertheless indirectly related, via the geometry of the device, since the suspension dimensions (i.e., choice of beam geometry to achieve a given stiffness) is strongly dependent on the material's Young's modulus.

Doubly Clamped Cantilever with Mass at Center

Most MEMS designed to sustain reasonable shock have a symmetrical design and spread load over at least two anchors, leading to an s-shaped beam deflection. So we shall redo the computation above, but now for a mass suspended by 2 anchors. The motivation is to compute the magnitude of the maximum vertical load (hence acceleration) that micromirrors such as the one in Figs. 4.26 and 4.27 can handle. The mirror in Fig. 4.26 is suspended by two poly-silicon springs from a gimbal, which is suspended by two poly-silicon springs from the frame.

Fig. 4.26 Polysilicon micromirror from Alcatel-Lucent. Two 2 μm wide serpentine springs attach the mirror to the gimbal, which is attached by two more 2 μm wide serpentine springs to the frame (outermost ring). Reprinted with permission of Alcatel-Lucent USA Inc.

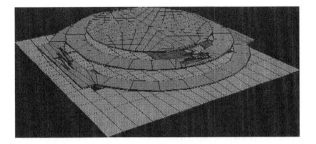

Fig. 4.27 FEM response to a vertical shock, showing the mirror and the gimbal rising up. Displacement is not to scale to emphasize motion. Reprinted with permission of Alcatel-Lucent USA Inc.

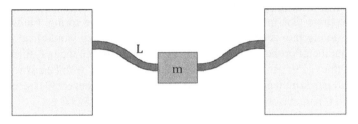

Fig. 4.28 Schematic cross-section of a symmetrically suspended micromirror

The simplified model of the mirror is shown in Fig. 4.28, where two beams of length L, thickness t, and width w are bent as the acceleration moves the mirror vertically.

The maximum moment for each of the two beams of length L for this doubly-clamped configuration, neglecting any bending of the mass, is:

$$M_{\max} = -\frac{F}{4}L$$
$$= -\frac{maL}{4}$$

The maximum stress is:

$$\sigma_{\max} = \frac{M_{\max}}{I}\frac{t}{2}$$
$$= -\frac{3}{2}\frac{maL}{wt^2}$$

So the critical acceleration for fracture for the 2-spring device in Fig. 4.28 is:

$$a_{\text{critical}}^{2-\text{springs}} = -\frac{2}{3}\frac{wt^2}{mL}\sigma_{\text{critical}}$$

For a device like the one in Fig. 4.26 with a gimbal (so 4 springs in all: 2 from frame to gimbal, 2 from gimbal to mirror), the maximum sustainable acceleration is roughly the same, assuming the gimbal does not deform, as the two springs

supporting the gimbal also bend, leading to twice the displacement, and hence the
same maximum strain in each spring as for without the gimbal.

$$a_{\text{critical}}^{4-\text{springs, gimbaled}} = -\frac{2}{3}\frac{wt^2}{mL}\sigma_{\text{critical}} \qquad (4.5)$$

We shall now apply equation (4.5) to three micromirrors, one made from polysilicon, and two from SOI. Two of these designs have folded (serpentine) springs.
For these folded beams, we shall use in the calculations for L the length of one
segment. This leads to a slight overestimate of maximum sustainable shock, but
is more accurate than using the full unfolded length of the spring. Finite Element
Modeling taking into account the exact shape of the spring would lead to a more
accurate number. For the micromirror in Fig. 4.26, for which the polysilicon mirror
is 2.6 μm thick, 250 μm radius, and hence a mass of 1 μg, with beams with $t=$ 2.6
μm, w=2.0 μm, arm length of each serpentine beam segment of 50 μm, and a total
length of 400 μm, we obtain $a_{\text{critical}} = 1.8 \times 10^5$ m/s^2 = 18,000 g.
  These micro mirrors were shocked tested in a specially built setup, which allowed
shock testing up to 25,000 g. For shocks in the vertical (pushing the mirror "up"),
failures were seen starting near 5,000 g with many mirrors surviving higher shock
levels [136]. In view of the approximations in the model above and the variations
in fracture strength for brittle material, there is reasonable agreement between the
experimental data on shock susceptibility and the prediction.
  Now consider the SOI mirror from Alcatel-Lucent in Fig. 4.29, based on 100
μm long, 3 μm thick and 1.2 μm wide torsion beams [135]. The mirror mass is
4 μg, and applying equation (4.5) leads to a predicted $a_{\text{critical}} = 1.1 \times 10^5$ ms^{-2} =
11,000 g.
  The SOI mirror from Alcatel-Lucent in Fig. 4.30 is a design that is very tolerant
to residual stress and to vertical forces during release because it is based on serpentine beams. The folded suspension beams segments are 60 μm long, 1.1 μm wide

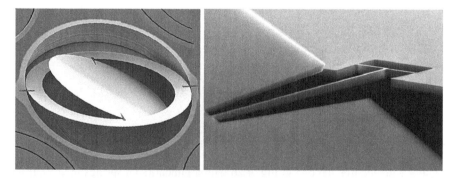

Fig. 4.29 875 μm diameter SOI mirror, suspended by 1.2 μm wide, 3 μm thick torsion beams
[135]. Reprinted with permission Copyright 2003 IEEE

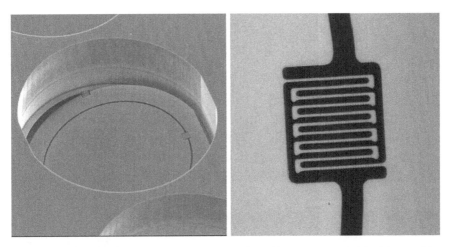

Fig. 4.30 (*Left*) 875 μm diameter SOI mirror from Alcatel-Lucent suspended by 1.1 μm wide, 5 μm thick serpentine beams (operating in torsion). Reference [28] reprinted with permission Copyright 2003 IEEE. (*Right*) Optical micrograph of the silicon suspension spring (high-shock resistance design). Reprinted with permission of Alcatel-Lucent USA Inc.

and 5 μm thick [28]. The mirror mass is 7 μg. Applying equation (4.5) leads to a predicted $a_{critical} = 2.6 \times 10^5 = 26{,}000$ g.

When these SOI mirrors were shocked tested some of them failed by spring fracture at only 200 g, though those in Fig. 4.30 only fail at shocks well above 1000 g. To explain this, and to explain how the design was improved in order to exceed the thousand g shock levels required for Telcordia Generic Requirements for Single-Mode Fiber Optic Switches (GR-1073, discussed in more detail in Chapter 6), one must leave our simplistic static calculation and take the details of actual geometry into account, namely including:

- Stress concentration
- Surfaces coming into contact, stiction (i.e., failure may not be due to fracture)
- Ringing and different modes, and hence the duration of the shock pulse
- Shocks coming from arbitrary directions (the model above only considered the vertical piston mode)

Figure 4.31 shows the evolution of the serpentine spring design that allowed progressing from failures occurring at 200 g due to stress concentration at sharp corners to a design that withstands shocks of greater than 1000 g on all three axes. 73 mirrors with the final design were shocked tested repeatedly on all three axes and no failures were observed. It is worth emphasizing that failures need not be due to fracture suspension beams but can also be due to stiction of parts that are normally would not come into contact or to delamination. With careful engineering, even large MEMS structures can be made highly shock resistant.

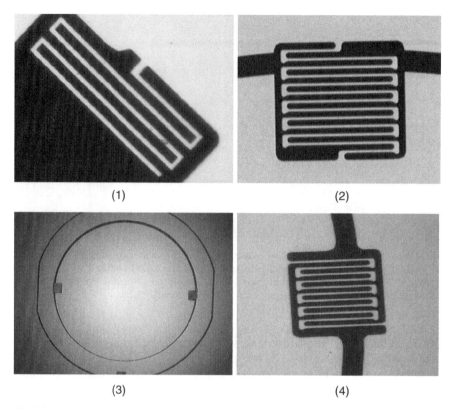

Fig. 4.31 Changes in spring design on Lucent Technologies SOI micromirrors to go from 200 g to >>1000 g shock survivability. Reprinted with permission of Alcatel-Lucent USA Inc. 1. First generation of SOI mirrors failed at 200 g. Failure mode: cracking of the spring at the 90° corner. The image shows the fractured beam after vertical shock. 2. So the sharp corner where stress was concentrated were eliminated, see image above. Failure then occurred at 400 g, *not by fracture*, but by stiction from lateral motion of the mirror. 3. Mirror with spring as in (2) after lateral shock: the mirror "slid" under the gimbal, and stuck. Failure mode is stiction. 4. "Rotated" serpentine beam providing enhanced lateral stiffness for the same torsional stiffness as in (3), but allowing the mirror to survive repeated shocks in all axes for shock levels greater than 1000 g, 0.5 ms half-sine

4.2.2.5 Conclusions on Shock

We summarize some general conclusions on making MEMS more shock resistant.

- By virtue of the small mass of MEMS devices, shocks of a few g are easy to accommodate.
- Shocks of up to 1000 g can readily be dealt with by spring design (avoiding stress concentration, symmetrical designs, . . .)
- Shock of 10,000 g require more careful design (of MEMS but also of attachment and package)
- No fatigue has been observed from shocks.

- "Stoppers" are a widely used technique to mechanically limit motion of beams. This approach is very effective as it minimizes displacement and kinetic energy, but stiction can be an issue.
- One must design the suspension to uniformly spread loads, for all 3 axis, and testing must be done on all three axes.
- To avoid current spikes that could damage the device, one must ensure that surfaces which might come into contact are at the same electrical potential
- Cleanliness is essential to avoid any particulates that could move and the two short-circuits or mechanical blockage.
- The die attach material must be carefully chosen, balancing strength with induced stress.
- Use the package to dissipate the shock load.
- Keep in mind that fracture is not the only failure mode (stiction can play a large role).

4.2.3 Vibration

In this section, we describe a general methodology to determine a lower limit on vibration level expected for failure as a function of frequency.[1] The objective is to estimate the vibration acceleration needed to bring parts into contact or to reach fracture stress, prior to vibration testing, in order to avoid surprises and to serve as a tool for the designer. Vibration testing is discussed in Chapter 6, from a qualification perspective and then presenting an example of vibration testing of a polysilicon MEMS micro-engines from Sandia.

The procedure to determine a lower limit on maximum safe vibration levels involves three steps.

1. List the possible failure modes due to vibration: i.e., list the displacement in 3 axes that could lead to stiction or short circuit or to fracture. For instance, for a SOI micromirror suspended 2 μm over the handle layer, with a 10 μm clearance around the periphery, a 2 μm motion in the -z axis could lead to stiction, and a 10 μm motion in the xy plane could lead to stiction.
2. Measure device response vs. frequency (in plane & out of plane) to obtain resonant frequencies and quality factor (damping). This will allow the dynamics of the device to be determined.
3. Use method below to generate plot of safe lower limit of vibration below which contact will not occur, and therefore below which failure is not expected. This plot is merely a lower limit, since contact does not automatically lead to stiction.

---

[1] This procedure was developed by Subramanian Sundaram at the EPFL.

Fig. 4.32 Forced vibration
model, with fixed support.
The MEMS mass is driven
directly (e.g., electrostatically
actuated mirror)

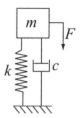

For the purpose of shock and vibration characterization, most MEMS devices
in air or in vacuum can be accurately modeled as a single degree of freedom mass
spring system, as in Fig. 4.32, with mass m, spring constant k, damping constant c,
driving force F. The displacement x can be written as a function of frequency ω and
first resonance mode ω_{res} as:

$$x(w) = \frac{x(w=0)}{\sqrt{\left(1 - \left(\frac{\omega}{\omega_{res}}\right)^2\right)^2 + \left(2\xi\frac{\omega}{\omega_{res}}\right)^2}} \qquad (4.6)$$

where $\xi = \dfrac{c}{4m\omega_{res}}$.

Now consider instead Fig. 4.33, which shows a model for the case of a mass
excited at the base (i.e., using a shaker). The parameter of interest for vibration
testing is the motion x of the mass relative to the motion y base can be written as:

$$z(t) = x(t) - y(t)$$

Using the coordinate system depicted in Fig. 4.1 and summing all the forces that
act on the mass, the equation of motion for the system can be written as

$$m\ddot{x} + c\dot{x} + kx = c\dot{y} + ky \qquad (4.7)$$

Fig. 4.33 Support vibration
model, where the support is
being vibrated. The MEMS
mass motion $x(t)$ is due the
support motion $y(t)$. The
motion of interest is the
relative motion x–y

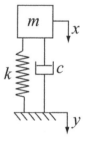

Solving the equation for the steady state and rearranging the terms, we can write a displacement function as a ratio of the amplitudes of the motion of the mass and the excitation base in the form:

$$\frac{z}{y} = \frac{\left(\frac{\omega}{\omega_{res}}\right)^2}{\sqrt{\left(1 - \left(\frac{\omega}{\omega_{res}}\right)^2\right)^2 + \left(2\xi\frac{\omega}{\omega_{res}}\right)^2}} \tag{4.8}$$

To proceed, one needs ω_{res}, $\xi = 1/(2Q)$ and $x\,(\omega=0)$. This must be obtained experimentally, for instance from Laser Doppler vibrometer data or from stroboscopic video microscopy. Figure 4.34 is an example of amplitude vs. frequency for an in-plane mode of an SOI electrostatic MEMS device taken with a Veeco Wyko NT1100 using stroboscopic video microscopy. From the fit to (4.6), ω_{res}, Q and $x(\omega = 0)$ are extracted.

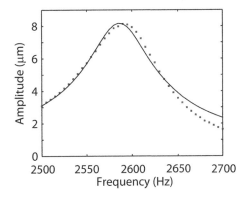

Fig. 4.34 Frequency response of an MEMS device (data points) and fit to (4.6), giving $\xi = 0.0135$, and $\omega_{res} = 2.59$ kHz

Using (4.8) one can then determine the input acceleration at the shaker necessary to produce failure as a function of the excitation frequency, having previously determined the displacement that could led to stiction or fracture. We shall call this the critical acceleration $a_{crit}(f)$, below which vibration cannot lead to failure (for the given mechanical mode). It is possible that the device operates reliably above acceleration $a_{crit}(f)$, since contact between two parts does not necessarily imply stiction or fracture.

$$|a_{crit}| = y\omega^2 = z\omega_n^2 \sqrt{\left(1 - \left(\frac{\omega}{\omega_n}\right)^2\right)^2 + \left(2\zeta\frac{\omega}{\omega_n}\right)^2}$$

An example is shown in Fig. 4.35; the input acceleration has a minimum for an excitation frequency equal to the resonant frequency of the device. For this example, the data from Fig. 4.34 was used, and failure was assumed to occur at a displacement

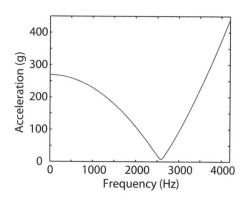

Fig. 4.35 Computed minimum critical acceleration a_{crit} vs. frequency to obtain 10 μm motion, using the parameters extracted from Fig. 4.34. Vibration levels below the curve will not lead to failure

of 10 μm. One obtains a critical acceleration of 7.6 g at resonance for that in-plane resonant frequency.

This procedure is to be repeated for the different vibration-induced possible contacting surface for a device. Generally, only a two or three failure modes need to be considered as one will dominate for each axis. The use of a Laser Doppler vibrometer and stroboscopic video microscopy is discussed in more detail in Sections 5.5 and 6.3.5.

This approach assumes the MEMS device is unpowered during the vibration. If the device is being actuated while being vibrated, the critical acceleration for failure could be greatly reduced for instance because snap-in could occur, and because of electrostatic spring softening.

4.2.4 Creep

4.2.4.1 Introduction

Creep is the time-dependent increase in strain in a solid at constant temperature and stress, i.e., creep is plastic deformation under applied strain. By definition, creep occurs only in ductile materials, so what follows applies to metal thin films, not to silicon, (except at temperatures above 600°C). There is one exception for MEMS: polysilicon which has been galvanically corroded by HF during release has shown creep [29, 30]; this can be avoided by proper etch procedure, as described in Section 4.4.2 and references therein.

Creep was introduced in Chapter 2 (Section 2.5.1), in the context of the case study of plastic deformation of the TI DM mirrors, and the deformation mechanism map of homologous temperature versus normalized shear stress was introduced. Creep is generally a consequence of dislocation motion, and depends on: Temperature, Stress (intrinsic, thermal, or applied) and Time. It is expressed as $d\epsilon/dt$, where ϵ is the strain and t the time. Temperature plays a key role in atomic diffusion and dislocation mobility. The Homologous temperature (ratio of operating to melting temperature T_{melt}) provides a good guide as to dislocation mobility. The following three temperature regimes are often used:

$0 < T < 0.3\ T_{melt}$: no creep observed

$0.3\ T_{melt} < T < 0.9\ T_{melt}$: dislocation motion leading to creep

$0.9\ T_{melt} < T < T_{melt}$: diffusion creep (nearly liquid flow)

A commonly used criterion is that creep is appreciable (relatively fast deformation, accelerating with time) for temperatures larger than $0.5\ T_{melt}$, and is slower (and with a rate that is decreasing) below this threshold. The threshold of $0.5\ T_{melt}$ serves merely as a guide to give an order of magnitude of what a suitable operating temperature will be. The value of $0.5\ T_{melt}$ is given for a few materials in Table 4.3. Due to their low melting temperature, solders generally exhibit creep near room temperature. One should also note that the device temperature can be significantly higher than ambient, for instance in RF MEMS switches carrying a few 100 mW of power the membrane can reach 200°C [31], and in projector applications spatial light modulators near the incandescent light source can exceed 100°C if no precautions are taken.

Table 4.3 Temperature at which $T_{homologous}$=0.5 for several materials

Material	T_c=0.5 T_M (in Kelvin)
60% Sn – 40% Pb (solder)	–45°C
Pb	27°C
Al and Al alloys	190°C
Ti	700°C
Si (brittle)	570°C
W	1600°C

4.2.4.2 Reducing Creep in MEMS

Increasing the creep resistance of a MEMS device can follow 3 paths: (1) reduce the operating temperature), (2) reduce the applied stress levels, (3) change the material. We shall see in the following example from TI that all three approaches were needed to reach the desired operating life.

A complete change in material may not be possible from a process point of view, but often only a small change in film composition is needed. The purpose of a material composition change is to block dislocation motion, which can be achieved by pinning of dislocations by solute atoms, impeding dislocation motion by short–range order, and increasing dislocation density to tangle them (work hardening). Defects and grain boundaries can trap or pin dislocations. This is done most easily by introducing tiny particles of a second phase into a crystal lattice, for instance Fe_3C in steel, or Al_2Cu in Al.

Creep is currently an important failure mode for RF MEMS switches, and was initially an important failure mode for Texas Instruments' DMD micromirrors. TI refers to creep as "hinge memory", because creep leads mirrors that have been titled for extended periods in one direction more than the other to exhibit a small residual tilt. The residual tilt increase the required drive voltage to tilt in the "less-used" direction, eventually leading to pixel failure when the voltage margin is used up.

Factors that TI identified as contributing to hinge memory were temperature and duty cycle [32]. By replacing the original Al torsion bars by another material (an aluminum alloy with increased creep resistance or that allowed for lower maximum stress), they were able to obtain a fivefold increase in lifetime. To obtain a further factor of 5 improvement in lifetime, TI did not reduce the creep rate (i.e., did not change the peak stress or the spring material), but implemented a different electrical waveform that allowed reliable operation with larger residual tilt. Finally, to reach the reliability level need for a consumer product, they implemented a thermal management (a system-level fix as well as a packaging-level fix) to keep the mirror array temperature below 45°C under normal conditions, operating only 7–10°C above ambient temperature. With these three changes, they predict array lifetime in excess of 100,000 h for failure due to hinge memory, see Fig. 4.36 [32].

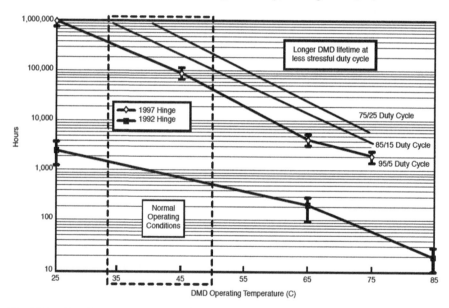

Fig. 4.36 DMD lifetime estimate for failures from hinge memory (creep), for two hinge generations (1992, 1997) as a function of temperature, and duty cycle, with 95% being an accelerated test condition. From [32] reprinted with permission Copyright 1998 IEEE

Another example of reducing creep in MEMS comes from IMEC, regarding the choice of metal for membranes in RF switches [33, 31]. They compared 5 different Al alloys for creep resistance ($Al_{98.3}Cu_{1.7}$, $Al_{99.7}V_{0.2}Pd_{0.1}$, $Al_{99.6}Cu_{0.4}$, $Al_{99.6}Cu_{0.4}$, $Al_{93.5}Cu_{4.4}Mg_{1.5}Mn_{0.6}$), and correlated the microstructure, determined from SEM and TEM images (principally precipitate size and distribution, as well as direct imaging of pinned dislocation), with the creep rate. They observed much lower creep in films with small (nm size) and highly dense particles, which were most effective at pinning dislocation and hence slowing creep. Up to 110°C, they showed that dislocation glide limited by obstacles (the precipitates) is the dominant creep mechanism.

The IMEC groups also showed that annealing and quenching can change the grain interior by changes the precipitate size and dispersion, thus allowing fine tuning of creep resistance by a hardening process, as used in the processing of many metals for macroscopic use.

The choice of film depends on the desired resistivity (slightly higher for the Al alloys), stress as deposited (higher for the alloys), thermal budget for annealing, material availability, and acceptable creep rate.

4.2.4.3 Metal Films on Silicon MEMS

Silicon is in principle immune to creep, as a brittle material. For optical MEMS devices, or for improved electrical conduction, silicon is often coated with a thin metal film. It is essential for long-term creep-resistance to ensure there is no metal on the suspension element or elements under high stress, as the metal can creep, thus applying stress to the underlying silicon, making it look like the silicon MEMS is creeping. One must distinguish between metal on silicon beams (where it is generally only needed for the case of thermal actuators) and metal on optical reflectors, where the metal is always needed.

A silicon beam a few μm thick that is beam covered with few hundred nm of Al can lead to appreciable creep (or hinge memory). By way of example, an early generation of Lucent Technologies SOI micromirrors had the 5 μm thick silicon suspension beams coated with approximately 100 nm of Al. Following accelerated testing (high applied stress) for 5 days at 85°C, a residual tilt of 0.1° was observed. By contrast, after 2 months of accelerated testing, micromirrors whose beams did not have the metal coating showed no residual tilt.

Reflective coatings on micromirrors cannot be eliminated, as they are required for the desired optical performance. This type of micromirror is presented in Section 5.3.1, with a discussion of curvature. A big challenge these mirrors pose is during annealing (e.g., during packaging, die attach, wirebonding) when the mirror deforms due to the CTE mismatch, but when the metal then plastically deforms and leads to a flat reflector at the bonding temperature, often of order 150°C. When cooled to the ambient temperature, the mirror can have a significant curvature due to the CTE mismatch between silicon and metal. This curvature will slowly decrease if the metal creeps at room temperature, though this process can take many months. Single-sided metallization will thus generally relax to a flatter state, as the metal creeps but not the underlying silicon. If both sides have been metalized to create a symmetrical and thus initially flatter mirror, there is however the potential for an uncontrolled increase in mirror curvature if the creep occurs differently for the two metals films, which is quite likely given that they were not deposited under identical conditions.

One can of course obtain flatter metalized mirrors by making the silicon thicker, but this decreases the resonance frequency, increasing response time and susceptibility to mechanical shock, or leads to very high drive voltages, with associated dielectric breakdown and dielectric charging issues.

4.2.4.4 Conclusions on Creep

Silicon MEMS are not affected by creep below roughly 600°C, as long as there is no metal on the suspension. This holds for SiN_x and SiO_x flexures too.

For metal MEMS, creep can be an important failure mode, which can be mitigated by reducing the applied stress (by geometry of material change), reducing operating temperature (better heatsink, different package), or a change of material (either to a brittle materials, or more commonly to an alloy with much higher creep resistance).

4.2.5 Fatigue

Since MEMS have moving parts, fatigue was initially thought to be an important failure mode, especially for parts requiring many operation cycles. Fatigue has not turned out to be a lifetime limiting factor in any commercial MEMS device. For silicon, the material most commonly used in MEMS, fatigue occurs only for applied stresses greater than half the single-cycle fracture strength, i.e., at stress levels close to fracture, and thus any reasonable design will not have stress levels sufficiently high for fatigue to be relevant. For metal MEMS, fatigue can occur at lower relative stress levels, but is generally much less problematic than creep (plastic deformation) or other failure modes such as charging.

In view of the extensive research carried out on silicon MEMS, it is now well known how to avoid fatigue (by controlling the maximum stress and the relative humidity). The topic will therefore be briefly addressed for silicon. Metal MEMS, because of their lower melting point, are more susceptible to fatigue.

4.2.5.1 Introduction to Fatigue in Brittle and Ductile Materials

Fatigue is the cycle-dependent decrease in yield strength, i.e., a slow crack growth leading to failure due to a periodically applied stress. The maximum stress at each cycle is below the single cycle fracture strength, yet at each cycle of alternating stress, the crack grows, reducing the strength of the material, and eventually leading to failure.

The key concept for fatigue is that fluctuating loads can lead to failure when monotonic loads do not. Some materials, such as steel, display an endurance limit: a critical stress level below which failure does not occur regardless of number of cycles. Aluminum and polymers do not show such a limit. Fatigue data is often plotted as a stress-life (S/N) curve, plotting the maximum applied cyclic stress vs. the number of cycles to failure (see Fig. 4.38 for data on micromachined silicon).

Ductile materials (e.g., most metals) and brittle materials (e.g., silicon, ceramics) exhibit very different fatigue behavior. For ductile materials, fatigue generally occurs due to plastic deformation at the crack tip involving dislocation motion, leading to alternating blunting and sharpening of an existing crack tip. Fatigue can therefore occur over a large range of stresses. Brittle materials do not plastically deform at ambient temperatures as they lack dislocation mobility, so for brittle materials the crack progresses by cycle dependent degradation of the toughness of the material in the wake of the crack, and thus fatigue only occurs for stress levels near the yield strength [34].

For macroscopic materials the mechanisms of fatigue crack propagation are well summarized in reference [35]. For MEMS devices, with their large surface to volume ratio, and critical dimensions comparable to grain size, the surface and microstructure play an essential role in fatigue properties.

4.2.5.2 How to Measure Fatigue in MEMS

In view of the small size of MEMS devices, standard test structures commonly used on macro-scale samples for fatigue measurements cannot be used. For silicon (poly-crystalline and single crystal) the most widely used test device is the one first proposed by Van Arsdell and Brown at MIT [36], and shown in Fig. 4.37. It consists of a free-standing silicon proof mass, roughly triangular with 300 μm sides, suspended by a single notched beam. Two comb drives are used, one to electrostatically drive the mass at resonance (in plane, roughly 20–40 kHz), and the other to capacitively measure displacement in order to determine the resonance frequency. The stress is maximum at the notch, which can be pre-cracked with a nano-indenter. As the crack grows at every cycle, the resonant frequency decreases. By measuring the evolution of the resonant frequency with time and environmental conditions the crack growth can be determined. These resonators allow crack growth rates of down to 10^{-12} m/s to be measured [36]. See also in Chapter 5 for the use of laser Doppler vibrometry to measure resonance frequency.

4.2.5.3 Silicon MEMS

Silicon is a prototypical brittle material, in which fatigue has never been observed in air at room temperature for bulk samples. Dislocation activity not observed at low homologous temperatures ($T_{ambient}$ / T_{melt} <0.3), and there is no evidence of extrinsic toughening mechanisms, such as grain bridging, nor of stress corrosion cracking (environmentally induced cracking). So fatigue was not expected in silicon in air room temperature.

Yet, as reported by Van Arsdell et al [36], Muhlstein et al. [37, 38, 27], and Kahn et al. [39, 40] polysilicon films from both the MUMPS and SUMMiT^{TM} processes show failures due to fatigue after 10^6 to 10^{12} cycles when operated in ambient air at stresses as low as 1/2 of single-cycle fracture strength, see Fig. 4.38 for stress-life (S/N) curves from several authors.

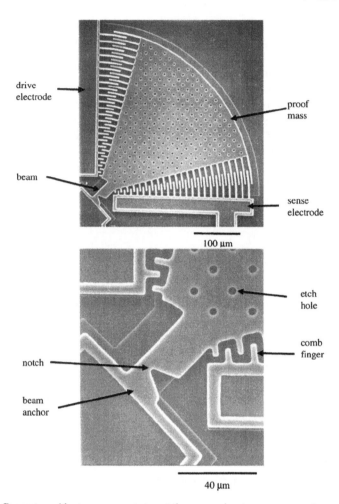

Fig. 4.37 Resonator with stress concentrator at the suspension to measure crack growth due to fatigue [36]. Reprinted with permission Copyright 1999 IEEE

Relative humidity plays a key role in fatigue lifetime for silicon. Muhlstein et al. [37, 38] and Alsem [41] have shown that fatigue in silicon MEMS is due to a reaction-layer fatigue process, occurring in two steps. First, at locations where cyclic stresses are maximum, the post-release oxide is thickened. Second, this oxide undergoes moisture-assisted cracking, leading to sub-critical cracks growth. Once fresh Si is exposed at the crack tip, it expands upon oxidizing, further driving the crack growth at each cycle. The presence of moisture is required for this crack growth.

The data in Fig. 4.38 shows clearly the effect of humidity on the fatigue life of polysilicon MEMS: in high vacuum, no fatigue is observed at cyclic stresses of over 4 GPa after 10^{11} cycles. For ambient air (roughly 30–40% RH) fatigue is clearly

Fig. 4.38 Combined maximum cyclic stress-lifetime (*S/N*) data for polysilicon MUMPS and SUMMiT V^{TM} devices, different types of devices are tested in ambient air (25°C, 30–40% RH), high relative humidity (25°C, 95% RH), and very high vacuum (25°C, <2.10^{-7} mbar). Reference [41] reprinted with permission Copyright 2006 American Institute of Physics

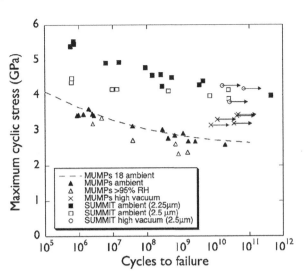

seen for large stresses, and at 95% RH, the stress leading to fatigue failure is lower than for ambient air.

Even single crystal silicon shows high-cycle fatigue in air [27], with the tested samples showing lives from 10^5 to 10^{11} cycles before failure for stress amplitudes from 4 to 10 GPa.

While the research groups discussed above have clearly shown the existence of fatigue in silicon MEMS, and experimentally plotted a S/N curve, one should note obtaining this data required carefully designed test structures that can only reach the required stress levels at resonance.

As a general rule, if either (a) the maximum cyclic stress is less than 20% of single-cycle fracture and if humidity is not controlled, or (b) maximum cyclic stress is less than 40% of single-cycle fracture and the device is hermetically packaged in an ultra-dry ambient, high-cycle fatigue of silicon parts will not occur.

Since silicon is a brittle material and thus exhibit a range of fracture strength, the conservative MEMS designer will tend to limit the maximum designed stress to below 20% of yield strength, thus avoiding at the same time fatigue. Single-crystal and poly-silicon micromirrors from Lucent Technologies for instance have not shown any fatigue effects after over 10^{10} cycles in ambient air, because the maximum stress levels in the suspension beams is only a few percent of yield strength, which is also why they survive the very large externally imposed displacement in Fig. 4.2.

4.2.5.4 Metals

Metal films are used in a variety of commercial MEMS, notably Texas Instruments DMD mirror array. Since metals are ductile and have mobile dislocations at ambient temperature, plastic deformation and hence fatigue can be a life-limiting factor. Care must be taken to remain in a "safe" section of the S/N curve.

As for silicon, special test structures were devised to test MEMS metal films, in view of the length-scale dependence of material properties, which for Al films was fond to be very important. The two main metals are LIGA electroplated Nickel, and Aluminum.

TI mirrors have a lifetime estimate of over 100,000 h with no pixel failure [42]. At a mirror modulation frequency of 7 kHz, each micromirror needs to switch about 2.5×10^{12} cycles. TI therefore extensively studied fatigue, but do not report observing any failures, running an array of 307,000 micromirrors up to 10^{12} cycles with no failures, corresponding to 2×10^{18} total mirror movements [32].

TI had expected to observe fatigue failures, based on standard models for Al bulk samples. In macroscopic samples, the initial crack forms at the surface grain boundaries due to dislocation pile-up, leading to a crack when the dislocation density is high enough. The crack then grows by further dislocation motion at the tip, with the associated plastic deformation. The Al alloy films used by Ti for its suspension are however only one grain thick (approximately 100 nm). M. Douglas proposes that the two free surface of each grain are effective at relieving stresses due to dislocations, preventing the accumulation of a high enough dislocation density to form fatigue cracks [32].

Electroplated Nickel, used in the LIGA process, has been studied for fatigue by several groups (e.g., [43, 44, 45, 46]), and the properties for thin films is found to be similar to that of macroscopic bulk annealed samples, with an endurance limit of order 200 MPa, which is two orders of magnitude less than for silicon. Figure 4.39 is a S/N plot for electroplated Nickel.

Since fatigue in metals is associated with plastic deformation, the same precautions to minimize creep are also effective at minimizing fatigue related failures: (1) re-engineering the suspension to minimize stress levels (avoiding stress

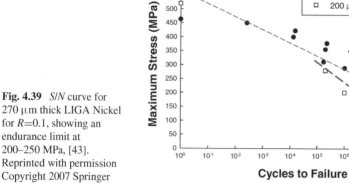

Fig. 4.39 *S/N* curve for 270 μm thick LIGA Nickel for *R*=0.1, showing an endurance limit at 200–250 MPa, [43]. Reprinted with permission Copyright 2007 Springer

concentration); (2) reducing the operating temperature; (3) choosing a more creep-resistant material, such as an alloy rather than a pure metal.

4.3 Electrical Failure Modes

Because MEMS devices, unlike integrated circuits, have moving parts, their failures are often thought to be mechanical in origin. This is an overly simplistic assumption, which might have been true in the early days of MEMS, but is no longer the case. As design rules for MEMS have matured, mechanical failures have become increasingly uncommon, and electrical failures can play a large role in MEMS lifetime.

4.3.1 Charging in MEMS

4.3.1.1 Introduction to Dielectric Charging

Dielectric charging is only a concern for MEMS devices that are sensitive to charge, namely principally electrostatically driven or sensed MEMS devices. MEMS using electromagnetic or thermal actuation or sensing principles are generally insensitive to dielectric charging.

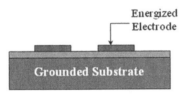

Fig. 4.40 Schematic cross-section of a MEMS electrode, consisting of a substrate (e.g., Silicon), a dielectric (e.g., SiN_x or SiO_x) and electrodes (e.g., Al or Si)

Electrostatic MEMS devices often require high operating voltages (50–200 V) applied across small gaps (0.1 to a few μm), resulting in electric fields of order 10^8 V/m across the dielectric. A simplified cross-section is shown in Fig. 4.40. The high fields across both the bulk and along the surface of the dielectric can give rise to charge injection that contributes to several possible failure modes. These bulk and surface leakage currents, and associated trapped charge, are a strong function of voltage, temperature and relative humidity. In MEMS devices, the dielectrics often also serve a structural role, and their stress must be carefully tailored. This is general done at the expense of electrical properties (e.g., reduced breakdown field, increased trap density).

For fixed applied electrode voltages, the electrostatic force on a MEMS actuator or sensor is assumed to be constant in time. This situation however only holds in the ideal scenario where the dielectric contains no mobile charges or charge traps, so that all electric fields are uniquely determined by the voltages applied to the electrodes. Charging of the dielectrics in MEMS structures gives rise to undesired and difficult to predict time-varying electrostatic forces which are a serious performance issue for a wide range of electrostatically driven or sensed MEMS devices including microphones, displays, micromirrors, and RF switches.

The failure modes related to dielectric charging are drift in applied electrostatic force as a function of time, leading to a gradual shift in actuation voltage (e.g., calibration change), or a gradual change in rest or actuated position, or a gradual shift in release voltage. For a micromirror, this leads to a time dependent tilt angle, which can lead to large insertion loss or snap-down of the mirror and associated stiction concerns. For an RF switch, this can lead to a device which eventually is stuck in the "down" state, when the trapped charge provides a larger holding force than the restoring force of the suspension, or a device stuck in the "up" state, when the trapped charge screens the applied voltage (see case study in Chapter 2 for MEMtronics RF switches). Minimizing dielectric charging is also important so that MEMS devices can be operated in "open loop", i.e., without the need for complex and possibly bulky or power-hungry feedback electronics and sensors.

Sensors such as some accelerometers and gyros that use a capacitive read-out scheme can also be very susceptible to charging, and the stability of the output is greatly enhanced by the same techniques that minimize actuator drift. Designs that eliminate charging are usually radiation hard, as charges in dielectrics due to ionizing radiation will not affect device performance. We shall return to this point in the section on radiation effects.

There are a number of very effective techniques to eliminate or mitigate charging, but they often entail reliability trade-offs, as will be discussed below.

Origin of Charging

When a DC bias is applied across a dielectric, charge carriers from the electrodes can be injected into various charge traps in the bulk or on the surface of the dielectric. Leakage currents can occur on the surface of the dielectric between electrodes on the dielectric held at different potentials, or through the bulk of the dielectric when there is a potential drop across the dielectric. This charge injection leads to the buildup of a quasi-static charge on the surface or in the bulk of the dielectric. In addition, mobile ions (such as Na^+) can migrate on the surface of the dielectric, and this situation is significantly worsened by the presence of any adsorbed water layers on the surface, as occurs in the presence of humidity.

The charging and discharging times to fill or empty the traps can be different by orders of magnitude, and are typically much longer than the mechanical response of the MEMS device (typically minutes or hours to charge the traps, vs. milliseconds response of the MEMS device). Unless the dielectrics are suitably electrically shielded from the actuator, the time-dependent charge on or in the dielectric gives rise to a time-dependent electrostatic force on the actuator, whose equilibrium position or force then changes with time. This "drift" of the actuator position is of electrical, not mechanical, origin (i.e., it is not due to plastic deformation of the supporting springs).

Because of the high fields ($\sim 10^8$ V/m) applied across dielectrics in electrostatically actuated MEMS devices, conduction is typically non-ohmic, and is dominated by conduction via traps in the dielectric, and by charge injection and tunneling. For applications where the relative dielectric constant is not an important parameter (e.g., when the dielectric is used for electrical insulation rather than to make a capacitor), the most common materials in MEMS are silicon nitride and silicon oxide.

It is not clear what happens on an atomic scale for charging and dielectrics. Different behavior (charge/discharge time constants, trap densities) is observed for slightly different deposition or growth techniques (such as CVD, PECVD) and a strong dependence is seen on annealing conditions and film stoichiometry.

The two main conduction mechanisms through those dielectrics are the Frenkel-Poole (FP) and Fowler-Nordheim (FN) models. FP conduction describes charge transport dominated by traps, and so very accurately models conduction in Si-rich SiN films commonly used in surface micromachining. FN conduction, which does not rely on defects or traps, describes tunneling of electrons from the electrode conduction band into the dielectric conduction band through part of the potential barrier at the conductor-dielectric interface. The FN model is most appropriate for conduction through silicon oxides.

While the details of the charge accumulation process are not well understood, it is generally accepted that the total trapped charge is the integral of the injected current. This is why it is important to understand the leakage current.

The Frenkel-Poole current is proportional to:

$$j_{\text{F--P}} \propto \exp\left[\frac{q\left(\beta\sqrt{E} - \phi\right)}{k_{\text{B}}T}\right] \qquad (4.9)$$

where $\beta = (q/\pi\, n^2)^{1/2}$, q is the electronic charge, n the index of refraction, E the electric field across the dielectric, and ϕ the activation energy for conductance mechanism.

The Fowler-Nordheim current scales as:

$$j_{\text{FN}} \propto \frac{E^2}{\Phi_B}\exp\left[-\frac{8\pi\sqrt{2qm}(\Phi_B)^{3/2}}{3hE}\right] \qquad (4.10)$$

Where Φ_B is the potential barrier for electron injection into the oxide (for Si/SiO$_2$ interface $\Phi_B \approx 3.2$ eV), m the electron mass, and h is Planck's constant.

As an illustration that Si-rich SiN$_x$ follows very accurately the Frenkel Poole model, Fig. 4.41 shows leakage current measured through 0.6 μm thick silicon nitride (from a MUMPS run at Cronos in 2002, now MEMSCAP). By measuring the leakage through different films one can get an order of magnitude estimation of the amount of charging.

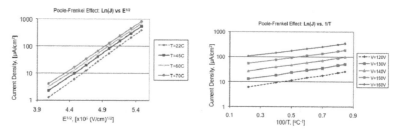

Fig. 4.41 Scaling of leakage current through the dielectric as a function of inverse temperature and square root of the electric field for Cronos (now MEMSCAP) SiN$_x$, showing clear Poole-Frenkel scaling. (Data courtesy of A. Gasparyan). Reprinted with permission of Alcatel-Lucent USA Inc.

4.3.1.2 Mitigation of Charging Effects

In this section, we focus on two devices to illustrate the effect of charging due to high applied electric fields: silicon micromirrors with SiN$_x$ and SiO$_x$ dielectric under the actuation electrodes, and metal RF capacitive switches with SiN$_x$ and high-k dielectrics. In the section on radiation effects, we shall focus mostly on polysilicon accelerometers as examples devices failing due to charge buildup from ionizing radiation.

As is generally the case for reliability issues, there is a trade-off between performance and reliability. Many of the solutions below, in particular those related to redesigning the device, come at the cost of increased fabrication complexity or packaging cost. For instance charging can be greatly reduced by hermetic packaging, but hermetic packaging can cost of up to $1000 per package for large arrays of micromirrors. It should be noted that traps will charge and discharge faster at higher temperature and that therefore charging can be less of a problem at higher temperatures. Since most failure modes are accelerated by temperature, following an Arrhenius model, heating the device to influence dielectric charging is not a generally acceptable solution because of the lifetime penalty associate with higher temperature operation.

We list some possible solutions to charging, then delve into more detail for the effect of geometry, charge dissipation layers and carefully engineered voltage levels.

There are a number of documented ways to solve or minimize the "charging" problem, including:

- Bipolar AC drive voltage
- Geometry changes to
  - minimize area of exposed dielectric, or pattern the dielectric
  - shield movable parts (sense mass, actuators) from electric fields due to trapped charge.
  - Selectively remove dielectric to avoid charging

- Charge Dissipation Layers to remove surface charge and provide shielding
- Change dielectric or change composition to reduce amount of trapped charge or decrease discharging time constants (e.g, SiO_x instead of SiN_x)
- Reduce electric fields (e.g., thicker dielectric or with higher dielectric constant, redesigned springs to operate at lower voltages)
- Optimized drive voltage (multi-level: one to actuate, one to hold), or charge monitoring
- Control of packaging ambient to minimize humidity and contaminants

Using a bipolar ac rather than dc voltage drive seems at first like the perfect simple solution, for instance as proposed by Reid and Webster [47]. It does indeed greatly reduce charging effects, but does not completely eliminate it, due to different time constant for filing and emptying traps of different polarity and of different types (surface, bulk, etc.). Since AC actuation requires more complex drive electronics and has significantly higher power dissipation, other approaches are often preferred. De Groot et al. provide a good overview in [48].

4.3.1.3 Geometry Changes

As a concrete example, and following closely reference [49], let us consider the SOI micromirrors developed by Lucent Technologies in 2000–2002, for which a schematic cross-section of a MEMS micromirror shown in Fig. 4.42. The bottom wafer ("electrode wafer") consists of a Si substrate covered by a dielectric (here SiO_x) on which electrodes (Al or poly-silicon) are patterned (over one of more wire routing layers). The small black dots represent trapped charges. A polyimide spacer is patterned on top of the electrode wafer, and an SOI wafer is flip-chip'ed onto the spacer. The micro-mirrors and supporting springs are etched out of the 5 µm thick Si layer in the SOI wafer. Applying a voltage to one or more electrodes tilts the mirror. Figure 4.43 is an SEM micrograph of one such mirror.

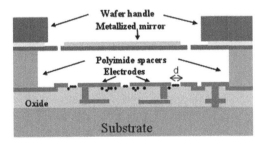

Fig. 4.42 Schematic cross-section of an electrostatically driven MEMS micromirror. Mirror and electrode wafers are fabricated separately and then assembled. The black dots between and under electrodes represent trapped charge in the dielectric, as well as slowly moving mobile charges on the surface of the dielectric. The substrate is grounded while the electrodes can be grounded or held at a fixed potential. Reference [49] reprinted with permission Copyright 2004 IEEE

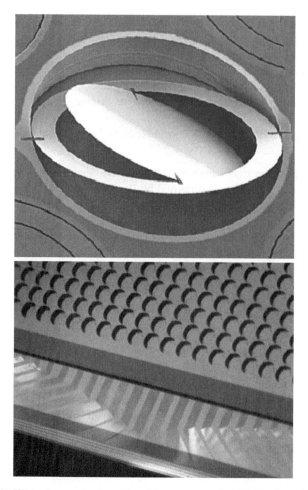

Fig. 4.43 (*left*) SEM micrograph of an Alcatel-Lucent two-axis MEMS micromirror fabricated from 5 μm thick single crystal silicon. The mirror diameter is 875 μm. The cross-section of this mirror is schematically illustrated in Fig. 4.42. Reference [135] reprinted with permission Copyright 2003 IEEE (*right*) optical micrograph of the assembled device, the wire routing is clearly visible on the lower part of the bottom chip. Reprinted with permission of Alcatel-Lucent USA Inc.

The advantage of this 2 chip approach for studying charging is that one has direct access to the electrodes and dielectric prior to bonding, allowing for more anti-charging techniques to be tried than for surface micromachined MEMS.

Once can mitigate the effects of dielectric charging by controlling the electrode and dielectric geometry: principally the width of the gaps with exposed dielectric between electrodes, the thickness of the electrodes, and selective etching of the dielectric. Approaches to minimize drift by changing the electrode geometry include creating overhanging electrodes to shield the actuator from the dielectric (though this may present a fabrication challenge and decrease the breakdown voltage).

The width of the exposed dielectric (i.e., the gap d between neighboring electrodes in Fig. 4.42) plays two roles: first, the larger the exposed area of dielectric there is under the actuator, the larger the electrostatic force that surface charge can exert on the actuator. This is a strong motivation for narrow gaps. Second, it is known that the dynamics of charge transport on the surface of dielectrics can be characterized using a diffusion model [49, 50]. This model suggests that to first order the saturation time t_s scales with gap size d and surface diffusion coefficient D as $t_s \sim d^2/D$. For silicon oxide, D is of order 10^{-11} cm^2/s. Narrower gaps not only reduce the area of exposed oxide thus decreasing the magnitude of charging induced drift, but also shorten the saturation time. Therefore, small gaps between electrics are helpful for reducing the adverse effect of charging on mirror tilt angle stability. Note however that minimizing anodic oxidation (see Section 4.4.2) and increasing in-plane breakdown voltage (see Section 4.3.2) calls for larger gaps: a careful consideration of packaging and operating voltages and environment is required before deciding on the ideal gap size for a given application.

In Fig. 4.44 the drift (due to dielectric charging) in micromirror tilt angle is plotted for two Lucent Technologies MEMS micromirrors of similar geometry but with different gaps between electrodes. For a 10 μm gap, over 0.1 degree drift are observed in 15 h (with a saturation time is of order 100 h). For a 2 μm gap only 10 millidegree of drift are observed, with full saturation after 1 h [1]. Reducing the gap from 10 to 2 microns should to first order reduce the saturation time by 25 times, not out of line with what was observed.

A more effective and radical solution is to simply remove the dielectric from regions where the field from trapped charge in the dielectric can exert an electrostatic force on the MEMS device, as reported in [49]. In Fig. 4.42, the trapped

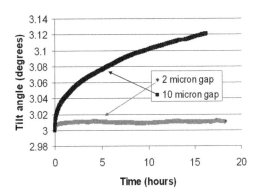

Fig. 4.44 Tilt angle vs. time for two Lucent micromirror test vehicles. The same dc voltage was applied to both at time $t=0$, but one mirror had a 10 μm gap between the actuation electrodes, and the other a 2 μm gap. The smaller gap show much shorter saturation time, and much less charging (hence less change in tilt angle with time). Reference [49] reprinted with permission Copyright 2004 IEEE

charge under the electrodes is not of concern, since it is shielded by the electrodes. The trapped charge between the electrodes however will give rise to undesired electrostatic forces. Starting with an electrode design with 2 μm wide gaps between electrodes, the mirror is shielded from the dielectric by undercutting the oxide in the gaps around the electrodes with a wet etch that stops on the underlying polysilicon shield layer, as shown schematically in Fig. 4.45. After the undercut, the mirror "sees" only conductive surfaces, thus eliminating charging induced tilt angle drift. Charge may build up in the remaining dielectric, but because of the geometry, these charges cannot give rise to any electric field at the mirror. Figure 4.46 is an SEM micrograph of an electrode chip where the oxide has been etched away: the undercut is clearly visible.

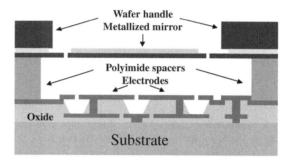

Fig. 4.45 Schematic cross-section of a MEMS micromirror device similar to the one shown in Fig. 4.42, but with the oxide selectively etched under the electrodes so that the mirror is fully electrically shielded from any trapped charge in the remaining oxide. Figure 4.46 is an SEM micrograph of the bottom chip of such a device after the isotropic oxide etch. Reference [49] reprinted with permission Copyright 2004 IEEE

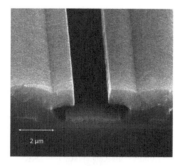

Fig. 4.46 SEM micrographs of the gap between 2-level poly-Si electrodes, with the exposed oxide between levels removed by wet etch. The electrodes (overhanging structures) are spaced by 2 μm. Rather than leaving oxide between the electrodes, the bottom of the gap between electrodes is covered by a grounded strip of poly-Si (running up the center of the image). The electrical potential of all surfaces is well defined, and the MEMS mirror is shielded from trapped charge in any remaining oxide. Drift in tilt angle due to dielectric charging is completely eliminated. Reference [49] reprinted with permission Copyright 2004 IEEE

This technique was found to be highly effective and led to micro mirrors with drift of less than 10 millidegree per day when held at a 5-degree tilt (i.e, below the measurement accuracy, and at level where drift in mirror tilt angle added less than 0.1 dB to the loss of the optical cross-connect in which an array of 256 or 1196 such micromirrors were used). Etching away the dielectric is a very effective solution, but care must be taken not to overetch the dielectric, which might lead to lower breakdown voltages, and not to damage the electrodes, which typically are made from poly-silicon in order to survive the oxide etch. This solution would be very challenging to implement with Al electrodes because of their susceptibility to attack by HF, but is ideal for use with poly-Si electrodes when 2 or more levels are available.

4.3.1.4 Charge Dissipation Layers

Following [49], there may be cases where it is not feasible to etch away the dielectric as shown in Fig. 4.45, for instance when fabricating electrodes on top of CMOS circuits, or due to common limitations of the process flow, or when using a multi-user or standardized foundry process. In that case increasing the conductivity of the dielectric can be an effective means to control charge build-up in the dielectric.

A well-known solution to the charging problem is to deposit or grow a thin conductive layer on top of the dielectric in order both to bleed off surface charge and to screen bulk charge from the reflector. This Charge Dissipation Layer (CDL) must not contain charge traps, and must be a good enough conductor to efficiently drain charge and provide electrostatic screening, while not being so conductive as to short out the electrodes by drawing too much current. The CDL typically consists of a thin film of a poor conductor such as a doped oxide. Lithium Niobate modulators have a similar charging problem to MEMS (though no moving parts). For instance, US Patent # 5,949,944 describes a CDL for $LiNbO_3$ modulators.

Figure 4.47 is a plot of tilt angle drift for two identical Lucent Technologies micromirror devices, except that the electrodes of one device were coated with

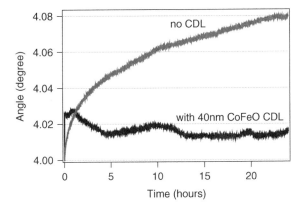

Fig. 4.47 Tilt angle drift for two Lucent Technologies SOI mirrors, on with and one without CDL, showing how the CDL effectively mitigates the effect of dielectric charging. Reference [49] reprinted with permission Copyright 2004 IEEE

40 nm of CoFe$_2$O$_4$ [49]. The Co-Fe-O CDL reduces both the magnitude and time constant of charging related drift by a factor of more than 10. The major advantage of Co-Fe-O is also its main potential problem: the conductivity the Co-Fe-O layer can be tuned over several orders of magnitude by annealing in oxidizing or reducing atmospheres. This allows for great flexibility in tuning of the CDL conductivity, allowing the Co-Fe-O films to be used for many different MEMS geometries and designs. The tunability however opens questions about the impact of high temperature packaging steps and about long-term stability of such coatings, which have not yet been studied, and would need to be carefully studied and acceleration factors identified before it could be used on a commercial product.

Rather than depositing a CDL over the dielectric, the dielectric material itself can act as a CDL if its electrical transport properties are suitable: dielectric materials with larger coefficient of surface diffusion and higher bulk mobility of charge carriers are less prone to static charge build up. However these more "conductive" dielectrics have lower breakdown electrical fields. This raises an interesting reliability vs. performance issue: extremely insulating dielectrics have larger breakdown fields, and thus offer higher protection against shorting through the dielectrics. Since electrostatically operated MEMS devices typically operate at voltages as high as 300 V, this is not a negligible issue. Slightly "leaky" dielectrics can make for devices where charging is much less of an issue, but lifetime may be limited by breakdown of the dielectric. For capacitive RF MEMS switches, Raytheon patented the approach of leaky SiN to control charging [52].

Figure 4.48 is a plot of tilt angle stability for 3 Lucent Technologies surface-micromachined mirrors of identical geometry, each fabricated on a different wafer by Cronos (now MEMSCAP). Each wafer has a slightly different composition of the Si-rich SiN$_x$ dielectric under the electrodes. Small changes to the composition

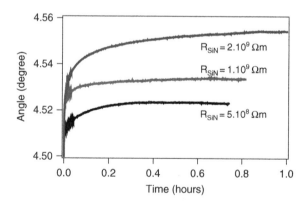

Fig. 4.48 Dependence of the charging-induced tilt angle drift of two Lucent Technologies surface micromachined mirrors on the resistivity of the underlying SiN dielectric layer: devices with lower resistivity dielectric films have less drift because charge is more readily drained away or screened. Reference [49] reprinted with permission Copyright 2004 IEEE

of the SiN change the film's conductivity, approximately 2×10^9 Ωm, 1×10^9 Ωm and 5×10^8 Ωm for the three samples. The higher the conductivity of the SiN, the smaller the angular drift, in line with the above argument that "leaky" dielectrics reduce dielectric charging (since charge leaks out), but at the expense of lowered electrical reliability.

A common model of dielectric breakdown is the "charge to failure" model [53]: the resistance of a dielectric in a large electric field remains very high even though electrons and holes are injected. These charge carriers damage the dielectric, creating more defects and charge traps. Once a critical amount of damage has been done, a conductive path is created through the dielectric, which has then broken down. So the higher the leakage current, the faster the critical charge will be reached. The thickness and conductivity of the dielectric must be carefully considered, trading off dielectric breakdown vs. charge mitigation.

4.3.1.5   Multi-Step Voltage Drive for RF MEMS Switches

This section addresses mostly charging in RF MEMS switches, very promising devices in terms of performance and integration, but whose commercialization is limited by it reliability issues. There are two main classes of RF MEMS switches: contact (or ohmic) and capacitive. We discuss here only the capacitive type, since the root cause of the main failure modes of such capacitive RF MEMS switches is dielectric charging.

The design and operation of RF MEMS switches are well described in [54] and [55], and was introduced in Chapter 2 for the third case study of a MEMtronics RF switch. Figure 4.49 is a schematic cross-section of the capacitive RF MEMS switch, which generally operates in two states. As seen in Fig. 4.50 it consists of a metallic bridge or membrane suspended above a conducting line or coplanar waveguide. There is a thin layer of high dielectric constant dielectric on the bottom trace to prevent a short-circuit when the membrane is deflected downwards electrostatically by applying a potential difference between the grounded membrane and the lower trace. In the "up" state (low capacitance state, C_{off}) with the top metal membrane

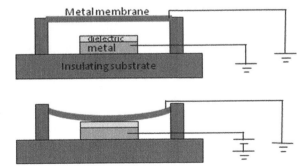

Fig. 4.49 Schematic cross-section of a capacitive RF MEMS switch, *top*: undeflected (no dc bias), bottom: snapped down (bias voltage larger than $V_{pull-in}$). In this geometry the metal trace serves both as DC actuation electrode and as RF transmission line

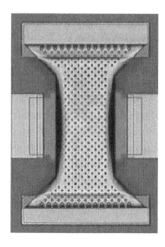

undeflected, the RF signal on the lower conductor propagates unaffected. In the
"down" state (high capacitance state, C_{on}) the top metal membrane is deflected
downwards by a DC bias on the central line, and signals in the lower conductor
in the gigahertz range are shunted to ground or reflected.

A performance metric for RF MEMS switches is the C_{on}/C_{off} ratio. C_{off} is defined
simply by the geometry and permittivity of the dielectric, but C_{on} depends both
on the dielectric constant of the dielectric, and on the flatness of the dielectric as
the membrane will never be in perfect contact with it and there is an effective air
gap remaining due to the roughness of the lower surface of the membrane and the
dielectric.

Capacitive RF MEMS switches might be expected to be reliable because, unlike
ohmic contact switches, is there is no direct metal to metal contact. They fail primarily due to parasitic charging of the dielectric which leads to a drift in both the
voltage required for actuation (pull-in voltage $V_{pull-in}$) and, more critically, the voltage below which the membrane return to the up position ($V_{pull-out}$), see Fig. 2.32.
$V_{pull-out}$ is much smaller than $V_{pull-in}$, typically 1–4 V compared to 30–80 V.

The dielectric is typically less than 300 nm thick (to achieve large C_{on} of order
1 pF), and capacitive switches usually require 30–80 V for actuation. In the down-
state therefore there is a very large electric field (greater than 10^8 V/m), and hence
charge will be injected into the dielectric. For silicon nitride films, which are often
preferred because they can be deposited at low temperature over metallization,
Frankel-Poole type charge injection is observed. For high-k materials, different
charge transport mechanisms may come into play.

Depending on the location of the trapped charge, the trapped charge can either
increase or decrease the voltage required to pull the membrane in. Bulk charge,
injected from the bottom electrode, reduces $V_{pull-in}$, while surface charge, on the
top of the dielectric, screens the applied voltage and increases $V_{pull-in}$. [56, 57].

Regardless of the polarity of the trapped charge, when the membrane is in the down-state the trapped charge provides an electrostatic holding force when the voltage is removed.

In normal operation, when the applied voltage is reduced below the pullout voltage, the membrane snaps back to its up position, assuming that the restoring force of the spring is larger than any stiction forces. However if $V_{\text{pull-in}}$ reduced below 0 V, the switch is stuck in the down position and hence has failed.

Fundamental work was done by Wibbeler et al. [58], modeling the shift in actuation voltage for a simple electrostatic parallel plate actuator due to trapped charge on and electrode. They also found that air discharge can be an important source of trapped charge.

Van Spengen et al. [59] measured and modeled charging in RF MEMS switches. They described in more detail the more complex effect of charged relation and provide a detailed model of the critical amount of charge required for failure of a capacitive MEMS switch, in particular taken into account the mechanical response speed of the switch which is generally much slower than the switching speed of the control signal. They find that there are critical positive and negative charge densities for failure, and that measurements are complicated by the slow discharging of the traps.

A key observation of van Spengen et al [59] is that the lifetime of an RF switch does not depend on the actuation frequency but on the total actuation time, as shown in Fig. 4.51. The key parameter is simply the total time spent in the downstate i.e., the time spent injecting charge into the dielectric, not the number of up/down cycles. The amount of accumulated charge, and hence lifetime, depends on the duty cycle and the applied voltage. The duty cycle is important because of the discharging that occurs when the actuation voltage is removed. Because fatigue had initially been

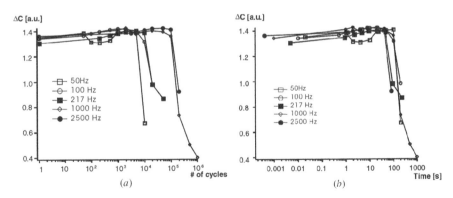

Fig. 4.51 Plot of the change in capacitance (essentially $C_{\text{on}}-C_{\text{off}}$) for RF MEMS test structures at IMEC, plotted as a function of the number of cycles (*left plot*), and as a function of total time in the down state (*right plot*). The device fails when the capacitance change per cycle decreases. It is obvious from these figures that the lifetime does not depend on the actuation frequency but rather on the total actuation time. Total actuation time is directly related to the total amount of charge injected in the dielectric. Reference [59] reprinted with permission Copyright 2004 IOP

a failure mode, and because MEMS manufacturers love reporting large numbers, much data on RF MEMS switches has been reported as a number of cycles to failure. It is important to note that it is really the time in the down state that is the key parameters that dictates lifetime.

Goldsmith et al. [60] have reported that the lifetime of a capacitive switch increases exponentially as the actuation voltage is decreased. This is shown in Fig. 4.52 for Goldsmith et al., and in Fig. 4.53 for van Spengen et al. The exponential dependence of lifetime on actuation voltage is reasonable in view of the exponential

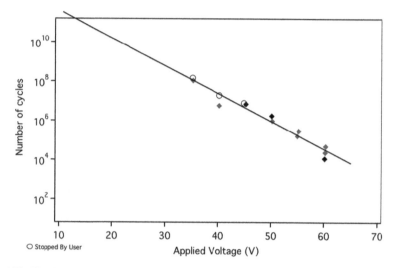

Fig. 4.52 Number of cycles to failure versus drive voltage for the switch of Goldsmith et al., showing an exponential decrease in lifetime vs. drive voltage. Adapted from [60] reprinted with permission Copyright 2001 IEEE

Fig. 4.53 Number of cycles to failure versus drive voltage for the switch of van Spengen et al., showing a roughly exponential decrease in lifetime vs. drive voltage. Adapted from [59] reprinted with permission Copyright 2004 IOP

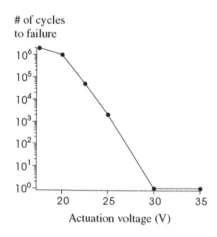

dependence on the square root of the voltage of the Frankel-Poole leakage current (see equation 3.1).

Van Spengen et al. developed a model for the time to failure:

$$t = -\tau e^{aV} \ln\left(\frac{\sigma_{\text{critical}}}{N_0 q} - 1\right)$$

where V is the drive voltage, σ_{critical} the critical charge density at failure, N_0 the total trap density in the dielectric, q the electron charge, and τ the charging time constant, a is a constant. N_0 and τ depend on the dielectric, and σ_{critical} depends on the switch design (geometry and materials mechanical properties). This model allows the lifetime to be predicted for different geometries and materials.

In view of the data above there are several possible solutions to increase the lifetime of RF MEMS switches that are failing due to dielectric charging.

- Use a dual voltage drive (unipolar, see Fig. 4.54)
- Use a bipolar drive (but power hungry)
- Design for lower voltage operation
- Change the dielectric to one with fewer trapped charge
- Modify the dielectric geometry (e.g., dielectric posts instead of a film) [56]
- Modify the electrode geometry (e.g., separate RF and DC electrodes)

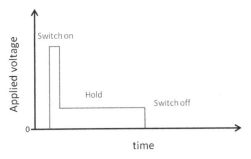

Fig. 4.54 Typical 2-level voltage rive scheme for an RF switch: first a high voltage pulse (30–80 V, lasting of order 1–100 ms) is applied to ensure the device has switched. Then a much lower holding voltage is all that is needed to keep the switch in the "down" state

All the changes imply a performance penalty, or a more complicated fabrication sequence. For instance, replacing the dielectric film with an array of dielectric posts as implemented by MEMtronics significantly reduces the total trapped charge, but also reduced the "on"-state capacitance, degrading switch performance.

Redesigning switch to work at a lower voltage is not trivial, though very appealing in view of the 10 fold increase in device lifetime (number of cycles, proportional to operating time for a given frequency) for every 5–7 V reduction in drive voltage [59]. More compliant springs reduce the actuation voltage but increase susceptibility to self-actuation, stiction, and vibration and shock. A larger electrode area also reduces the actuation voltage but takes up more chip real-estate, and requires larger membranes where residual stress and stress gradients can deform the membrane. A smaller gap reduces actuation voltage but also increases the off-state capacitance, and reduces maximum power handling due to self-actuation.

The generally implemented solution is to use a so-called dual pulse actuation waveform: apply a voltage larger than the pull-in voltage for the first 2 ms to actuate the device, then apply a much lower voltage (possibly only a few volts) to hold down the switch for the ms to hours or days that the switch is required to stay in the down position, as illustrated in Fig. 4.54. This solution allows keeping the stiff geometry with its associated performance benefits (large restoring force, high switching speed), but greatly minimizes the charge injection since the device is usually operating at the much lower hold voltage. The driver electronics are slightly more complicated, but this is not an important issue since the main cost driver in the control electronics for such switches is the charge pump required to get the 50–80 V needed for actuation.

One could also replace the dielectric with one that has fewer charge traps (e.g., SiO_2 instead of SiN_x). One must be careful with such a solution because the time constants may be significantly longer for oxides than for nitrides, but also because one pays a price in terms of possibly lower dielectric constant, and hence lower performance than due to lower capacitance. An important driver in this case is process compatibility: high-quality oxides are challenging to grow on metallization, nitrides are often chosen for ease of process integration. MEMtronics has reported switching from SiN to SiO, with an order of magnitude reduction in surface charging [56, 61].

MEMtronics obtained an important increase in lifetime (but at the expense of slightly reduced C_{on}/C_{off} ratio) by etching the sputtered SiO dielectric to form an array of pillars rather than a continuous film (see Fig. 2.37).

Finally one should mention that RF MEMS switches must generally be hermetically sealed to reach an acceptable lifetime, to avoid charge accumulation due to moisture and stiction. Humidity play an important role in surface charging, and has led to nearly all RF MEMS switches adopting a wafer-level packaging.

4.3.2 Electrical Breakdown and ESD

We shall distinguish between electrical breakdown though a solid dielectric (e.g., an insulator such as silicon nitride or silicon oxide film) and through a gas (e.g., arcing between neighboring electrodes). We shall also distinguish between electrical breakdown due to the sudden voltage and current pulse from an electrostatic discharge (ESD) event, and the lower but longer-lasting voltage from the normal drive signal for a MEMS device.

These distinctions lead to the organization of the section: first discussing breakdown through a gas, then though a solid, and finally the effect of ESD discharge on MEMS, for which the mechanical time constant is generally much longer than the pulse duration.

While the failure modes described here can be applied to any MEMS actuation or sensing principle, electrical breakdown occurs principally for the electrostatically operated MEMS devices, since electrostatic actuation often requires voltages of order 100 V.

4.3.2.1 Electrical Breakdown in a Gas for Micron-Scale Gaps

Electrodes for electrostatically operated MEMS are often spaced by as small gap as is possible given the fabrication technology. This is in particular true for comb drives were the electrostatic force scales inversely with the gap between comb fingers, with gaps of order 2 μm being common with actuation voltages up to 200 or 300 V.

One can distinguish several types of breakdown between two conductors [62] due to: (1) stressing of the electrode surface (also known as vacuum breakdown, related to vaporization of the electrode, i.e., vapor arc), (2) insulator breakdown (internal or external flashover), and (3) via the gas path (Paschen curve, i.e., avalanche ionization).

For MEMS, the Paschen curve (described below) was thought to provide a good description of breakdown. However, the limitation of the Paschen curve at micron-scale gaps at atmospheric pressure have become clear in the past few years [63–68]. The importance of the role of field emission and vapor arc have been demonstrated for gaps smaller than 10 microns, leading to the description of the "modified" Paschen curve, as discussed initially in [66], and illustrated in Fig. 4.55, plotting breakdown voltage vs. gap at fixed pressure of one atmosphere. The general conclusion has been that a maximum safe voltage is 300 V for gaps 4 microns or larger at a pressure of one atmosphere, and that the breakdown voltage decreases rapidly for smaller gaps. We return to these conclusions after a discussion of the Paschen curve.

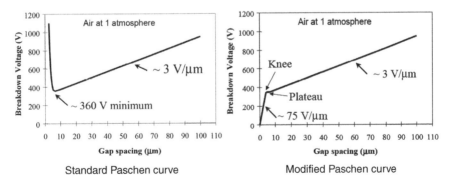

Fig. 4.55 (*left*) Theoretical Paschen curve in air at one atmosphere, plotting breakdown voltage vs. gap. (*Right*) modified Paschen curve, showing a reduction in breakdown voltage (absence of vacuum isolation) at μm-scale gaps. From [66] reprinted with permission Copyright 2003 Society of Photo Optical Instrumentation Engineers

In 1889, F. Paschen published a paper [69] which laid out what has become known as Paschen's Law. This law expresses the breakdown voltage V_{bd} of a dielectric gas as a function of the reduced variable $P_{red} = P \cdot d$, where P is the pressure and d is the gap between the two electrodes. His work was developed to understand the breakdown voltage between large metal plates at low pressure with macroscopic gaps.

Later work by J. Townsend [70] led to the understanding that the breakdown is an avalanche effect caused principally by the ionization of gas molecules by electrons accelerated by the electric field. If the electron gains sufficient energy between collisions to ionize gas atoms or gas molecules, then each collision gives rise to two electrons and an ion, allowing an avalanche effect eventually resulting in a spark. This avalanche can only occur when there are sufficient gas molecules between the electrodes, i.e., if the mean free path between collisions λ is much smaller than the distance d between electrodes: when $\lambda \ll d$. If the pressure is too low, or if the gap is too small, the avalanche breakdown (Townsend theory) cannot take place.

This absence of atoms or molecules is what gives the minimum in the Paschen curve. At large gaps or pressures, a linear relation breakdown voltage and electrode gap is found (reflecting the constant breakdown electric field of the gas), while at very small gaps one has a "vacuum isolation", where there are not enough gas atoms or molecules for the avalanche to occur. Another way to look at the breakdown is to consider the electron mean free path $\lambda_{electron}$ in the direction of the applied field. Like λ, $\lambda_{electron}$ scales inversely with the pressure (neglecting the Ramsauer effect) and so the product $P_{red} = P \cdot d$ is proportional to $d/\lambda_{electron}$, giving an indication of the number of collisions an electron undergoes when crossing the gap. The breakdown voltage V_{bd} then simply depends on the $P \cdot d$ product, all parameters except gap and pressure being fixed.

Later work led to the understanding that the Paschen curve also depends on secondary electrons emitted from the negative electrode when impacted by the positive ions. These electrons further accelerate the breakdown process. The secondary electron yield, γ depends on the cathode material. The Paschen curve can be obtained by computing the voltage required for the process of electron emission and multiplication to become self-sustaining [71]. One obtains:

$$V_{bd} = \frac{B \cdot P \cdot d}{\ln{(A \cdot P \cdot d)} - \ln{\left(\ln{\left(1 + \frac{1}{\gamma}\right)}\right)}} \qquad (4.11)$$

where A and B are properties of the gas, and γ is a property of the electrode material.

The Paschen curves were developed for macroscopic electrodes at operating pressures from a few Pa to one atmosphere. The generality of the scaling of V_{bd} with P_{red} led researchers to apply it to MEMS devices operating in a variety of gases at one atmosphere, for which a minimum breakdown voltage of order 360 V is predicted at a spacing of 8 microns (the exact voltage and minimum gap depend on the gas). This seemed like great news: regardless of design, the breakdown voltage in air at one atmosphere would be greater than 360 V.

Yet as reported in [63, 64, 67, 68], when the gaps are less than 10 μm for micro-machined structures operated at laboratory air at one atmosphere ($P_{red} < 1$ Pa·m), important deviations are seen from the Paschen curve. This regime is one where the mean free path is of order the gap, and thus where the Townsend breakdown cannot occur. Other types of breakdown are however possible. As presented for instance in [66] and in [68], field emission can become important at gaps order 5 microns, leading to a "modified" Paschen curve, which agrees with the "standard" Paschen curves at gaps larger than 10 microns ($P_{red} > 1$ Pa·m), exhibits a plateau of constant V_{bd} between 4 and 10 microns, and a linear drop in V_{bd} at lower gaps. Field emission can lead to local heating at micro-asperities on the surface the cathode, with in turn facilitates field evaporation of the cathode, leading to a cloud of atoms and ion in which an avalanche breakdown process can start [72]. Figure 4.56 shows data from Torres and Dhariwal for metal electrodes in air for gaps from 0.5 to 25 μm [73].

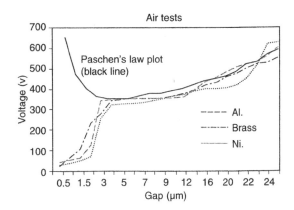

Fig. 4.56 Measured breakdown data for metal electrodes in air at one atmosphere, clearly showing the modified Paschen curve behavior [73]. With kind permission from Springer Science+Business Media: Microsystem Technologies, Volume 6, Number 1, November, 1999, pp. 6–10, Torres et al., Fig. 3

In addition to the nature of the gas (reflected in the constants A and B of equation (4.11)) and the nature of the electrode (in the form of parameter γ), relevant parameters that must be taken into account are the mean free path of gas atoms species, the surface roughness (which has a strong influence on the field emission), work function of the electrode, and the overall geometry of the electrodes, especially for planar geometries as found in MEMS and integrated circuits, which do not match the conditions of uniform electric field for which the Paschen curves were developed.

Different authors report in differences in the detailed behavior of breakdown voltage at small gaps. This is probably due to the dependence on electrode material, as mentioned above, and on surface cleanliness, which plays a large role. Nevertheless, one can roughly define a safe operating region for MEMS devices at one atmosphere, as illustrated by Strong et al. in Fig. 4.57. If one needs to operate above 300 V for gaps below 2–4 μm, careful testing must be done to ensure arcing or breakdown will not occur.

It is well known that different gases have different breakdown voltages, with the minimum in the breakdown voltage occurring at different reduced pressures (or, if operating at one atmosphere, at different gaps sizes) Of the commonly used gases,

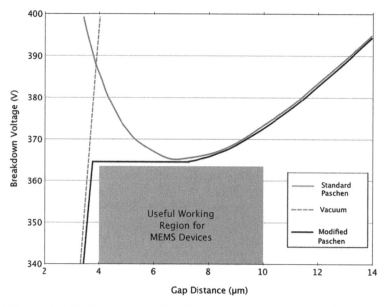

Fig. 4.57 Standard Paschen curve, modified Paschen curve, Vacuum breakdown curve, and safe operating region for MEMS for air at one atmosphere. Adapted from [68]. Reprinted with permission Copyright 2008 IOP

Neon and Helium have one of the lowest breakdown voltages, while Nitrogen has one of the highest. This is important when choosing the atmosphere for hermetically packaged MEMS, since a partial pressure of Helium is often used to aid in leak detection. Shea et al. showed that the breakdown voltage between polysilicon electrodes with gaps between 1 μm and 2 μm can be nearly 100 V higher in Nitrogen than in Argon [49], see Fig. 4.58. Ensuring sufficient margin to avoid arcing during device operation therefore requires careful selection of the packaging gas.

While most work on breakdown in small gaps serves to determine safe operating conditions for MEMS in air at one atmosphere, some MEMS devices operate at lower pressures, and thus have a different modified Paschen curve than the one

Fig. 4.58 Breakdown voltage vs. gap for polysilicon electrodes packaged under 1 atmosphere of argon and of nitrogen, showing the larger operating voltage possible with Nitrogen [49]. Reprinted with permission Copyright 2004 IEEE

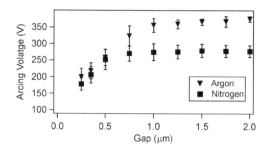

discussed above. The motivation may be micron-scale plasmas, in which case one generally seeks to minimize the breakdown voltage, or actuation in low pressures (e.g., MEMS scanner to be used on Mars) where on seeks to maximize the breakdown voltage. Carazzetti et al. [74] report on breakdowns in micron scale gaps for pressures between 1 and 800 mbar, and find for low pressures deviations from the Paschen curve for gaps even as large to 500 μm, which they explain principally by the planar geometry of typical interdicted his electrodes, that leads to the superposition of several Paschen curves and hence to a large flat region in the breakdown voltage vs. reduced pressure curve. They conclude that pressures well below 1 atmosphere, care must be taken when applying the Paschen formula to gaps on the 2–100 μm range, while for operation at 1 atmosphere passion behavior is observed for gaps larger than 10 μm.

4.3.2.2 Electrical Breakdown Across Solid Dielectrics

Electric breakdown across insulators is generally a two step process: (1) *wear out*: accumulating enough damage in dielectric to create a conductive path, followed by (2) Thermal damage from high current flow (*thermal runaway*).

In this simplified model, during the wear out phase, charge traps and defects accumulate in the dielectric and at the interface between conductor and dielectric. When the defect density reaches a critical level, the resistivity of the insulator plummets, a large current flows, and the device fails due to localized Joule heating.

Breakdown is extensively discussed in the semiconductor literature, for instance in ref [53]. The key concept is that of a critical charge to breakdown: Q_{BD}, with the assumption that the defect density increases linearly with current flowing through the insulator. It is for this reason that the leakage current through the dielectric is a key factor for dielectric breakdown. We discussed leakage current through dielectrics in the section on charging (Section 4.3.1.1), and refer the reader back to the section for discussion of leakage through oxide and nitrides.

The critical charge to breakdown concept is really shorthand for describing a critical defect density N_{bd}, above which the resistance of the device decreases markedly, leading to failure.

At first glance, for MEMS devices we are generally dealing with the same dielectrics as in the integrated circuit world. However there are two important differences: (1) the voltages may be higher than in typical CMOS, and (2) the insulator films used for MEMS often have compromised electrical properties to achieve better mechanical properties. The most common example is the silicon rich SiN film used in the poly-MUMPS (MEMSCAP Inc.) process. In order to minimize stress in the silicon nitride film, the film is made silicon rich. This leads to much higher leakage currents through the dielectric, and to lower breakdown voltages. This is well described in [49].

The reduction in breakdown voltage for SiN thin films is shown in Fig. 4.59 from stoichiometric $SiN_{1.33}$ to silicon-rich $SiN_{0.85}$ (based on experimental data in ref [75]).

Fig. 4.59 Breakdown
electric field for SiNx films of
different Silicon volume
fraction. Determined from
data in Fig. 2 and in Table 1
of [75]

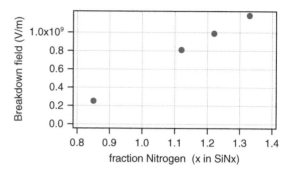

4.3.2.3 ESD and EOS

Electrostatic Discharge (ESD) is a sudden transfer of charge between two objects,
for example between a MEMS device and either a person handling the device, or a
piece of equipment. Electrical overstress (EOS) is very similar, but occurs at lower
voltage levels. Voltage pulses of kilovolts are typical for ESD, while EOS occurs
generally at tens to hundreds of volts. EOS often occurs because the device was
incorrectly wired, or incorrectly inserted into a socket; the voltage comes from a
power supply. ESD in contrast generally occurs during handling: a person walks on
a carpet in a dry environment easily charging up to 20 kV on a dry winter day and
then touches the pin of an electronic device, discharging through the device.

An ESD event typically has large voltages (kV) occurring in a short pulse (10 ns
risetime, 150 ns decay) and with a large current (>1 A). Tribo-electric effects are
an important cause of ESD as they result in static charge buildup: either charging
of a human, or friction between a chip and the plastic tube from which it is being
removed. While protection for CMOS and bipolar ICs is well established (see e.g.,
ref [76]), the literature is much sparser for MEMS. The first report of ESD on MEMS
was by Walraven et al. in 2000 on Sandia's micro-engines [77]. There have been a
number of reports on RF MEMS switches [78, 79], as well as a report on micro
mirror arrays [80]. The reports generally agreed that MEMS devices can be very
susceptible to ESD, and that suitable handling precautions must be taken.

An ESD event can cause both electrical and mechanical damage. Electrical damage includes: destroyed transistors, melted wires, weakened dielectric layers (hidden
damage), evaporated electrodes, charge accumulation in dielectrics. Mechanical
damage generally appear to be stiction failures, where surfaces not designed to come
into contact collide due to a sudden much static force possibly sticking using or
breaking.

The two most common models used to simulate ESD events are the human body
model (HBM) and the machine model (MM). The human body model simulates
the discharge occurring when a person handles a device, while the machine model
simulates a more rapid and severe electrostatic discharge from a charged machine,
fixture, or tool (i.e., a metallic connection rather than through a poorly conducting
finger). HBM simulates a person as a 100 pF capacitor that discharges through a
1.5 kΩ resistor, while MM simulates a machine as 200 pF capacitor discharging

this capacitor directly into the device being tested through a 500 nH inductor with no series resistor.

The rise time of the HBM ESD pulse, is between 5 and 9 ns, with a decay of order 150 ns. For a 400 V pulse, the current is 0.3 A. The rise time of the MM ESD pulse, is similar to HBM 6–8 ns, but the peak current at 400 V is 6 A, 20× greater than for HBM.

The timescales of these pulses are important because they must be compared to the typical resonant frequencies of MEMS devices. The reasoning here is similar to the one for shocks (see Section 4.2.2.2 in this chapter). The duration of the voltage pulse is much shorter than the typical mechanical response time of a MEMS device (at least for larger MEMS where stiction or impact is an issue; MEMS resonators can have GHz resonance frequencies, but are so stiff that electrostatically induced motion will not lead to failure), so the mechanical response of the MEMS device is that of an impulse: the device acts as if it had initial velocity given by the integral the acceleration pulse, and the dynamics of the device then simply follow from its resonance frequency and damping.

MEMS failure from ESD is often a combination of electrical and mechanical modes. Pure electrical failure has been reported, where the ESD pulse led to the breakdown of an insulator, or the evaporation of an electrical lead. Tazzoli et al [79] applied ESD pulses to RF switches and when applying the pulse along the signal line observed failures due to electromigration and electrically-open vias. When applying the ESD pulse between ground and the actuator, the observed sparking between the lines leading to failed leads. For both these cases the failures were not due to mechanical motion of the device.

Walraven et al. [77] report a number of failure modes, which involve a combination of electrical and mechanical failures following HBM and MM testing. Principally they observe that the ESD pulse leads to mechanical motion of a beam (comb finger in their case), which then comes into contact with a conductor a different potential, and fuses or "spot welds" (Fig. 4.60). They also report on comb fingers getting stuck on surfaces without any potential difference (stiction). So while the failure mode may appear to be stiction the root cause is the motion imparted by the voltage pulse. Failures occur at voltages less than twice the normal operating voltage.

A group from IMEC reported HBM and MM the ESD testing of arrays of polycrystalline Silicon Germanium micro mirrors [80]. The main failure mode if

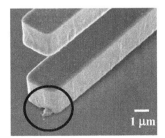

Fig. 4.60 Polysilicon comb finger welded to ground plane following 150 V ESD pulse (*black circle* around the weld) [77] Reprinted with permission Copyright 2000 Society of Photo Optical Instrumentation Engineers

they observe is irreversible pull-in, i.e., stiction or welding of the mirror in its completely tilted state. This occurs because under overvoltage conditions the mirror tilts and touches the actuating electrode, to which it then welds. This occurs at 40 V, which is only three times larger than the normal actuation voltage. ESD discharges at such low voltages can occur extremely easily, and would not be noticed by a person handling the device. Figure 4.61 illustrates an extreme case of mirrors being melted or blown off following an ESD pulse. This is explained by Joule heating due to the very large currents flowing through the mirrors. An interesting observation from this group is that HBM and MM discharges give very similar failure levels, which is never the case for CMOS devices. The high impedance of the electrostatic MEMS devices explains the very similar response to HBM and MM pulses. Such a device would require ESD protection being implemented in the drive circuitry, and will require very careful handling to avoid ESD damage.

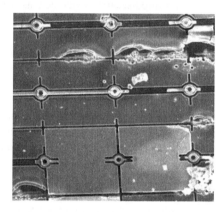

Fig. 4.61 SiGe micromirror array, showing extensive damage following ESD pulses up to 120 V. Mirror dimensions: $16 \times 16\ \mu m^2$ [80]. Reprinted with permission Copyright 2008 ESD Association

Short of implementing protection diodes, design changes can be made to MEMS devices to make them less sensitive to ESD. Electrical failures can be addressed by using a wider spacing between leads increases the voltage required for arcing, and wider leads that can tolerate larger currents before failing. Mechanical failures can be minimized by making the devices stiffer. Electromechanical failures can be mitigated by ensuring that moving parts only land on parts at the same potential, for instance an electrostatically actuated micromirror should have landing pads that are at the same potential as the mirror, and not land directly on the actuating electrodes.

These mitigation strategies all of course come at a cost, for instance higher actuation voltage for stiffer devices, or more complex electrode design. Trade-offs needs to be made between reliability, process complexity, and operating conditions.

4.3.3 Electromigration

Electromigration is the migration of metal atoms under an applied electric field. One must distinguish electrolytic (metal ions in solution) from solid-state (atom motion in a metal wire or trace due to electron momentum transfer) electromigration (EM) [81]. Electrolytic EM occurs generally on the printed-circuit board level when sufficient moisture is present to allow surface conduction between neighboring conductors by ionic transport, leading to failures by dendritic bridging. Solid state EM occurs in microfabricated wires, in which the momentum from electrons can cause atomic displacement of the conductors when the current density and temperature are sufficiently high, leading to voids and dendrites. Solid-state electromigration is an important failure mode for microelectronics in view of the very high current densities ($>10^{10}$ A/m^2) in IC circuits, and does not depend on ambient moisture.

We shall not cover electrolytic electromigration in this book, as it occurs mostly for silver under non-condensing conditions. It can occur for all metals when visible moisture is present, but we shall not consider this as a MEMS specific failure. In what follows we shall focus on solid-state electromigration, which has been a reliability problem in integrated circuits for over 40 years.

Because of the excellent thermal conductivity of silicon compared the substrates generally used for printed circuit boards or chip carriers, much higher current densities are possible in the thin-film interconnects of typical IC than in electrical motors or on printed circuits, with current densities of over 10^9 A/m^2 being common. The conventional physical explanation for electromigration is that the "electron-wind" force at high current densities transfers sufficient momentum to metal atoms to lead to a net mass transport.

Current-density driven electromigration results in momentum transfer to atoms with atom movement in the direction of the electron wind, which is opposite of the current flow. Thus voids will form where the where the electrons are injected, and hillock of metal atoms will accumulate where electrons are extracted. The grain structure plays a very important role in EM rates because diffusion of metal ions is roughly 6 orders of magnitude larger along grain boundaries than through bulk metal. Thus "bamboo" structure (interconnects only 1 grain wide) can carry significantly more current before showing failure due to EM than a wire several grains wide.

The accelerating factors for electromigration are: current density, temperature, and stress in the films. Temperature is accounted for in the typical Arrhenius manner, see Chapter 2. Black [82] is credited with the following equation:

$$\frac{1}{MTTF} = Bj^2 \exp\left[-\frac{E_a}{k_b T}\right]$$

where j is the current density, E_a the activation energy, k_b the Boltzmann constant, T the temperature and B a fitting parameter. For very high current densities the exponent of the current increases from 2 to larger values. The activation energy depends on metal species, as well as on grain size and purity. Typical values range from 0.4 to 1 eV, with 0.7 eV often used for typical Al-Cu IC interconnects. EM is well covered in IC reliability books, the reader is referred for instance to "Reliability and Failure of Electronic Materials and Devices" by M. Ohring for a more detailed discussion [53]. Much work has been reported on different metallizations to increase lifetime due to electromigration. Copper was added to Al to exceed the solid solution concentration for precipitation at the grain boundaries, (theta phase precipitates), and reduce surface energy. Ti-based underlayers can increase lifetime by an order of magnitude. When copper interconnects were introduced as a lower resistance alternative to Al-Cu, lifetime increased as copper can handle much higher current densities.

For integrated circuits such as microprocessors, EM can be a leading failure cause. For most MEMS devices, which do not have parts with high current densities, EM is not a critical issue. However for micromachined hotplates, used for instance in gas sensors, infrared emitters or membrane-type microreactors, EM can an important failure mode. It can also be observed in thermal actuators. Micro-hotplates consist of metal traces usually on a low-stress silicon nitride membrane to minimize thermal losses, and can operate at temperatures as high as 400°C. Electromigration occurs at the points of highest current density, or more correctly EM damage accumulates in areas of "flux divergence", where geometry or film thickness changes quickly. Figure 4.62 left is a top view of a micromachined hotplate [83], and Fig. 4.62 right is a SEM image following accelerated aging tests at 120 mW. The voiding due to EM is clearly seen. Lifetime can be increased by careful choice of metal and grain size.

Fig. 4.62 230 nm thick Pt heater on a 250 nm thick low-stress silicon nitride membrane. *Left*, optical micrograph. *Right*, SEM image after operation under accelerated conditions, showing voiding due to EM. Adapted from [83] reprinted with permission Copyright 2008 Elsevier

4.4 Environmental

The failure modes in this section involve degradation due to factors external to the MEMS device. Packaging plays a particularly crucial role when dealing with the interplay between environment and device. We shall address first the effect of radiation on MEMS, then anodic oxidation and galvanic corrosion of poly-silicon, as examples of MEMS-specific corrosion issues. Finally metal corrosion due to airborne humidity or ionic contaminants or atmospheric pollutants is covered as an issue affecting all microelectronic devices.

4.4.1 Radiation

This section is largely based on [91], to which the reader is referred for more details, in particular for different possible applications of MEMS on spacecraft (inertial sensing, propulsion, etc), also discussed in [133]. For reliability issues specific to operation in space, the reader is referred, for example, to [85], [86], and [132]. Qualification procedures for Space are discussed in Section 6.5.3.

Considerable effort has been expended over the past 50 years to devise techniques to test the suitability of electronics components for use in high-radiation environments, as well as design techniques to develop radiation tolerant electronics and optics. The physics of how different energetic particles interact with matter, the types of damage that are caused, and the influence on most electronic devices, optical components, and mechanical parts is well understood [87], and there exist well established test procedures for space applications, for instance [88] and [89]. Due in part to the relative immaturity of the MEMS field, but primarily due to the vast range of materials, technologies and applications that MEMS cover, there is no standard test procedure for the effect of radiation on MEMS, though there are some proposed approaches [85].

Even at the high end of space mission doses, the mechanical properties of silicon and metals are mostly unchanged (Young's modulus, yield strength not significantly affected). Silicon as a structural material can be viewed as intrinsically radiation hard. This makes most MEMS devices mechanically radiation tolerant by default. For MEMS devices operating on electrostatic principles, the main failure mode is the accumulation of charge in dielectric layers due to ionizing radiation. The trapped charge leads to device failure, for instance large changes in calibration of capacitive accelerometers, or device failure due to stiction initiated by electrostatic forces from the trapped charge. Of concern are also the drive/control electronics, which may need to be shielded or built with radiation-tolerant technologies.

4.4.1.1 Typical Doses for Space Applications

Consisting primarily of trapped electrons, trapped and solar protons, cosmic rays, and of bremsstrahlung (created when energetic particles strike the spacecraft), the space radiation environment is strongly time and position dependent. The dose received by the spacecraft (SC) thus depends by orders of magnitude on the SC orbit/trajectory, time of launch and duration of the mission. Although the radiation environment is complex, there exist excellent software tools (e.g. [90]) to model the dose and type of radiation a SC will encounter in its lifetime.

The radiation environment in space is complex, and is concisely described in [92–93]. Software models are available for the different types of radiation that can be encountered. Software packages (e.g., SPENVIS [90]) exist that combine these different models allowing rapid determination of the dose and type of radiation exposure for Earth orbits. Models also exist for deep space, but have less data to support them. The main types of radiation encountered near earth consist of:

- Trapped radiation: energetic electrons and protons magnetically trapped around the earth (Van Allen belts). They consist of electrons of energy up to a few MeV, and protons of up to several hundred MeV.
- Solar Energetic particles: mostly highly energetic protons, up to 300 MeV. The intensity varies greatly in time, especially the 11 year solar cycle, since the proton flux is associated with solar flares. UV and X-ray burst are also produced, as well as solar cosmic rays.
- Galactic cosmic-rays: continuous low flux of highly energetic (1 MeV to 1 GeV) particles, mostly protons, alpha particles, but also include heavy ions.
- Secondary radiation: radiation generated when the above radiation interacts with materials in the spacecraft, notably with shielding. Includes primarily electron-induced bremsstrahlung, but also secondary electrons, and other particles such as secondary neutrons.

The global effect of the many different types of radiation on components can be summarized by the quantity of energy deposited by the radiation. The SI unit is the Gray (1 J/kg), but the unit rad (1 rad = 10^{-2} Gray) is still in common use.

The energy deposited varies as a function of time and location of the SC. Accurate models can predict the quantity of energy deposited as a function of the trajectory. Table 4.4 gives approximate values of energy deposited in a component for a low Earth orbit (LEO) and for a geostationary orbit (GEO), without shielding and with shielding equivalent to 4 mm thickness of aluminum.

Table 4.4 Representative annual radiation doses for LEO and GEO orbits

Trajectory, shielding	Predominant particles	Dose deposited per year
LEO, outside SC	Trapped electrons	> 100 krad
LEO, 4 mm Al equivalent	Trapped protons	1 krad
GEO, outside SC	Trapped electrons	> 10,000 krad
GEO, 4 mm Al equivalent	Bremsstrahlung + solar protons	10 krad

Unshielded components obviously face a much harsher radiation environment that shielded ones. The spacecraft itself acts as a shield for components mounted internally (for a 5 tons satellite, this can be very significant shielding). The highest deposited dose is on solar panels and the external surfaces of the SC. On large space-crafts, it is unlikely that MEMS will be directly exposed to space so that much lower radiation values will be expected (with the exception of sun sensors and thermal control louvers [94]).

Space missions typically last several years, and operate in a radiation environment with dose rates of order 1 rad/h. Testing however must be done in hours or days (dose rates from 36 rad/h to 36 krad/h are commonly used for ^{60}Co irradiation).

Despite the complexity of the actual space radiation environment, accelerated radiation testing methods have been developed using mono-energetic particles whose relevance and suitability has been amply demonstrated for microelectronic devices. There is however no standard testing procedures established for MEMS, though studies are ongoing.

4.4.1.2 Damage Mechanisms

The effect of radiation on materials is well described in several books such as [87]. We briefly summarize in this section the main degradation processes and effects on different materials to serve as a foundation for a MEMS-centered analysis in the following section.

4.4.1.3 Degradation Processes

Energetic particles and photons cause damage by transferring energy to the materials they penetrate. The energy loss mechanisms are complex, but the type of damage can be classified in two consequences: (a) atomic displacement and (b) ionization. Figure 4.63 provides an overview of the effects that radiation can have on devices.

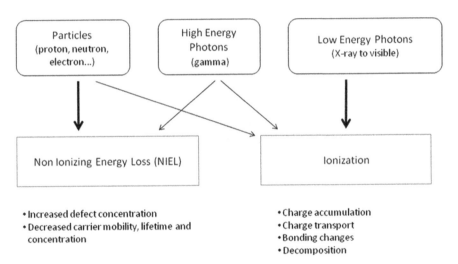

Fig. 4.63 Summary of radiation induced degradation effects, ignoring transient effects, adapted from Table 5.1 of [92]

Non-ionizing Radiation Loss (NIEL)

A fraction of the energy transferred to the target from energetic particles or even from photons results in the transfer of momentum to atomic nuclei, which can result in atoms being moved from their rest position in the lattice, leaving vacancies or defects behind. The process of atomic displacement is referred to as "bulk damage" [87]. Even photons of sufficient energy can give rise to this non-ionizing radiation loss (NIEL), or displacement damage, component of radiation. Displacement damage dose (DDD) is defined as displacement energy per unit mass, equal to NIEL time fluence.

The most relevant consequence of displacement damage for electronic devices is the reduction in minority carrier lifetime, the reduction of carrier mobility, and the removal of carriers (by interaction with defects). The damage caused by most particles is of the same general type.

Ionization

Most of the energy lost from radiation interacting with an absorber is ultimately converted to electron-hole pairs (the energy required is only 18 eV for SiO_2). Electrons and holes have very different mobilities. The electrons and holes increase the conductivity of the sample (even of insulators), and the holes can become trapped in insulators (SiO_x, SiN_x), leading to serious degradation of MOS and MEMS devices. This Total Ionizing radiation Dose (TID), defined as the Ionization energy per unit mass, leads to an accumulation of electrically active defects. The biasing of a sample is important because the electric field from the bias will drive the electrons and holes, and thus change the effect on the device of ionizing radiation. This will be seen below to be the driving factor in the radiation tolerance of MEMS.

Single Event Effects (SEE)

Single Event Effects are not a damage mechanism, but are an important consequence in microelectronics circuits of the effect of energetic particles, including Single Event Upsets and Single Event Latch-ups, which we do not discuss them further as they do not apply to MEMS devices (but do apply to the control/sense electronics).

4.4.1.4 Degradation Effects

The consequences of damage depend on whether the damage is due to atomic displacement or to ionization, whether the effects are transient or long-lived, and what type of material absorbed the radiation (we will distinguish between metal, semiconductor, and insulator). Additionally, one can also distinguish between changes in the mechanical vs. electrical properties of the materials.

Metals

There are no reports of important metal degradation by radiation in space [92]. In nuclear reactor cores the neutron fluxes are high enough to significantly reduce the mechanical strength of metals, or render them brittle. For space missions, metals are deemed to be radiation tolerant.

Semiconductors

Displacement damage leads to electrical and mechanical changes. The electrical changes are due to the change in minority carrier lifetime and concentration, which can have an important effect on p-n junctions (rectifiers and bipolar transistors, as well as solar cells). FET and MOS devices are much less sensitive to this effect.

Concerning mechanical changes, even at the high end of typical doses for space (Mrad), the amount of damage to silicon is rather small (defects, clusters), and the Young's modulus is not markedly changed. For electronics and packaging the effect can be ignored. For MEMS devices such as resonators, which are sensitive to ppm change in Young's modulus, further investigation is required.

Insulators

In optical materials displacement damage lead to color centers. For electronic or structural materials, displacement damage leads only to very small effects (compared to semiconductors) because dielectrics are typically glassy (amorphous), and there is thus no ordered lattice to disrupt with defects, clusters or dislocations. So the dielectric can retain its insulating properties even when a few atoms are displaced.

For dielectrics, ionizing radiation leads to both (1) direct charge injection from ionizing radiation, and (2) the creation of deeper traps and possibly more defects, thus making the dielectric even more susceptible to charging from non-radiation related sources. The influence of the trapped charge depends on the actuation scheme (electrostatic is much more sensitive), and on the geometry, such as the presence or absence of conductive shields to screen the trapped charge.

4.4.1.5 Review of Published Data on MEMS Radiation Tolerance

MEMS devices can operate on wide variety of physical principles for sensing and for actuation, the most common being electrostatic, thermal, magnetic, and piezoelectric. Other principles that are less widespread include chemical reactions, electrophoresis, and capillary force. The wide variety of materials and physical principles used make it difficult to make general statements about MEMS reliability and radiation sensitivity. Different sensing and actuation principles are shown below to very different in their radiation tolerance.

Few radiation tests have been performed on MEMS devices (less than twenty published papers), see Table 4.5 for an overview of minimum dose for failure and failure mode for different MEMS devices. Most radiation tests on MEMS have focused on the effects of radiation on the MEMS sensor or actuator, but have often been limited by failure of the electronics.

On the low-tolerance end, one finds that most electrostatically operated MEMS devices degrade between 30 and 100 krad, unless special steps are taken to shield

Table 4.5 Review of published sensitivity of MEMS devices to radiation. Failure modes are grouped below as mechanical related to displacement damage, mechanical related to ionization, and electrical due to charge trapping [84]

MEMS device	Actuation type	Minimum dose for failure	Radiation type	Failure mode	References
Analog device ADXL 150	Electrostatic (comb-drive)	27 krad (Si)	Co-60 γ	Not investigated	[97]
Analog device ADXL 150	Electrostatic (comb-drive)	Highly tolerant (no failures seen)	Infrared laser, 5.5 nJ (SEE)	Not investigated	[97]
Analog Devices ADXL 50	Electrostatic (comb-drive)	25 krad (Si)	Co-60 γ	Dielectric charging in device	[98]
Analog Devices ADXL 50	Electrostatic (comb-drive)	>50 krad (Si)	SEM localized e-beam 30 keV	Dielectric charging in device	[98]
Analog Devices ADXL 50	Electrostatic (comb-drive)	100 krad (Si)	5.5 MeV protons	Dielectric charging in device	[98]
Analog Devices ADXL 50	Electrostatic (comb-drive)	100 krad (Si)	155 MeV protons	Proton displacement in reference circuit	[98]
Analog Devices ADXL 50	Electrostatic (comb-drive)	20 krad (Si)	65 MeV protons and heavy ion	Dielectric charging in device	[99, 100]
Motorola XMMAS40G	Electrostatic	4 krad (Si)	Co-60 γ	Failure of CMOS readout circuit	[98]
Sandia microengines	Electrostatic (comb-drive)	1–100 Mrad (SiO_2) bias dependent	2 MeV protons, 5–25 keV electrons, 10 keV X-rays	Dielectric charging	[101]
Endevco accelerometer 7264B-500T	Piezoresistive	>30 Mrad	Co-60 γ	Trapped charge, depletion of minority carriers	[102]
Kulite pressure transducers XTE-190-25A	Piezoresistive	7 Mrad to >20 Mrad, sample dependent	Co-60 γ	Trapped charge, depletion of minority carriers	[102, 103]
DSTO / Analatom Si strain gauge	Piezoresistive	10^{15} protons/cm^2	3.5 MeV protons	decrease in carrier density and mobility (NIEL)	[104]
Sercalo 1×2 optical switch	Electrostatic (comb-drive)	>22.5 krad (Si)	Co-60 γ	No failures seen	[105]
Boston Micromachines Co Poly-Si Micromirrors array	Electrostatic (parallel-plate)	3 Mrad (Si)	Co-60 γ	No failure seen	[106]

Table 4.5 (continued)

MEMS device	Actuation type	Minimum dose for failure	Radiation type	Failure mode	References
Rockwell Scientific Co RF switch	Electrostatic (parallel-plate)	30 krad (GaAs)	Co-60 γ	Dielectric charging in device (strongly geometry dependent)	[107]
FBK-IRST ohmic RF switch	Electrostatic (parallel-plate)	10 Mrad (SiO$_2$) proton 1 Mrad (SiO$_2$) X-ray	2 MeV protons 10 keV X-ray	Both NIEL and ionizing damage	[108]
VTI SCA 600 accelerometer	Electrostatic (parallel plate)	50 krad (Si)	Co-60 γ	Not investigated	[97]
VTI SCA 600 accelerometer	Electrostatic (parallel plate)	Not quantified, but low	Infrared laser, 5.5 nJ (SEE)	due to latch-up in CMOS electronics	[97]
NASA /GSFC Microshutter array	Electrostatic & electromagnetic	10 to >200 krad (Si) depending on drive voltage	Co-60 γ at 60 K	Charge trapping (dielectric charging)	[109]
Polysilicon electrothermal actuator and bimorph cantilevers	Electrothermal and CTE mismatch	> 1 Mrad (Si)	Co-60 γ and 50 keV X-ray	No failure seen	[110]
Purdue wireless microdosimeter	Electrostatic (parallel plate)	Tested up to 650 Mrad	Co-60 γ TID	No failure seen	[111]

or remove the dielectric materials so as to render the device insensitive to charge build-up in dielectric layers. Tests on accelerometers and RF switches showed a marked change in calibration at doses above 30 krad [98, 100, 107]. Those failures were attributed to trapped charge in dielectric films. These doses are for unpackaged devices so that the sensor element is directly irradiated. Similar doses on packaged devices would lead to significantly less damage.

On the other extreme, micro-engines from Sandia National Labs in Albuquerque, NM, USA were reported to only change their behavior at doses of order 10 Mrad, in some cases over 1 Grad [101]. Those devices did contain dielectrics (SiO2 and SiN_x), but not in a geometry where charging could directly influence device operation.

Electrostatic MEMS Sensors and Actuators

For electrostatic MEMS devices the main failure mode at high radiation doses is the accumulation of charge in dielectric layers, which leads to failures as described in the earlier section on dielectric charging of this chapter. Therefore many of the same solutions to mitigate charging are applicable. While the failure may appear mechanical (e.g., a RF MEMS switch stuck in the actuated position, drift in tilt angle of an electrostatically actuated micromirror) the root cause is electrical. For a given device, total ionizing dose (TID) is the main radiation parameter that quantifies the amount of charging.

Accelerometers, in particular the monolithic comb-drive polysilicon devices manufactured by Analog Devices, which are readily commercially available, have been investigated for TID effects [98, 99, 97]. The devices operate by sensing the change in capacitance as a suspended proof mass moves in response to external accelerations. It is thus very sensitive to any static charge in exposed dielectrics, and Knudson et al. [99] showed the radiation-induced output voltage shift was due to charging of a dielectric under the proof mass. The devices tested under high energy proton and gamma-rays show degradation in the 50 krad range (ADXL 50 and ADXL 150). For similar devices where a conducting polysilicon film was placed over the dielectric (ADXL 04), thus effectively electrically shielding any trapped charge from the active device, no radiation induced degradation was observed up to a dose of 3 Mrad [99]. The XMMAS40G accelerometer from Motorola tested by Lee et al [98] failed after only 4 krad. It is proposed that the failure is due to failure of the CMOS output circuitry rather than the sensor element.

SOI bulk micromachined accelerometers from VTT, Finland, operating by measuring the capacitance between suspended parallel plates were subjected to gamma-rays, and failed at 50 krad [97]. The sensor was packaged with a readout-ASIC, which was found to latch-up at low doses of infrared laser pulses. It was not determined if the failure at 50 krad was due to the sensor or the ASIC. A non-monolithic approach (i.e., separate sensor and readout/control ASIC chips in one package) is an appealing approach to rapidly developing radiation toler-ant sensors, as it allows choosing a radiation-tolerant ASIC (an easier task since radiation hard CMOS technology is mature), and focusing the research solely on radiation-hardening the MEMS component.

Comb-drive actuators carefully designed with no exposed dielectric between or under moving parts (such as the Sercalo Microtechnology 1×2 optical switch [105] or the Sandia microengines [101]) have been shown to operate with no change after doses of more than 20 krad and 10 Mrad respectively.

Capacitive RF MEMS switches require a dielectric film to separate a fixed electrode from movable membrane. An RF switch from HRL Laboratories was successfully operated dynamically up to a dose of 1 Mrad [112]. RF switches from Rockwell Scientific Company reported in [107] showed no change in static characteristics at doses of up to 150 krad for design specifically developed to reduce dielectric charging. For a more conventional design, the device's calibration started to change at doses of 10 krad, although the device continued to operate after doses of 300 krad, but with an 80% increase in required drive voltage. The difference in dose required for degradation between the two devices is due to the (unspecified) different location of the dielectric layers. The configuration that is more radiation-tolerant has no dielectric between the moving parts.

Non-electrostatic MEMS Actuators

A piezoelectric mirror array developed by JPL and Pennsylvania State University based on PZT (lead-zirconate titanate) was functional up to 1 Mrad, but at 20 krad started exhibiting changes in mirror deflection compared to unirradiated samples, as well as an important increase in leakage current though the PZT [106]. The authors developed a model attributing the change in device characteristics to charge trapped in the PZT film.

Polysilicon thermal actuators and gold/polysilicon bimorph cantilevers were investigated by Caffey et al. [110] under ^{60}Co gamma-rays and 50 keV X-rays. No degradation of the devices was observed at 1 Mrad, the maximum dose used. This is in line with the understanding that electrothermal devices are for the most part insensitive to dielectric charging, as long as there is no exposed dielectric near the active element.

Piezoresistive Sensors

The radiation sensitivity of micromachined piezoresistive silicon accelerometers and pressure sensors are reported in [102, 104, 103, 113]. In all cases, an increase in resistance of the piezoresistive elements are observed. Marinaro et al. [104] find a nearly linear relation between the resistance of the piezoresistor in their single-crystal silicon strain gauge and the fluence of 3.5 MeV protons. They observed changes for fluences of the order of 10^{16} cm^{-2}, corresponding to roughly 10 years in MEO (Medium Earth Orbit). They attribute the increase in resistance to the NIEL component of the radiation, leading to majority charge removal due to displacement damage serving as trapping centers, and to a reduction in carrier mobility.

Holbert et al. [102] and McCready et al. [103] studied the response of piezoresistive MEMS accelerometers and pressure sensors to high gamma-ray doses and pulsed neutrons. They observed a gradual shift in output of Endevco 7264B-500T accelerometers with gamma-ray doses up to 73 Mrad, with no catastrophic failures, and were able to recalibrate the devices post-irradiation. Results were less consistent for Kulite XT-190-25A pressure transducers, with two devices failing suddenly at 7 and 25 Mrad, and four others still operating at after 20 Mrad, with a shift in output voltage. Holbert et al. [102] correlate the increase in resistance of the piezoresistors to the formation of trapped hole charges. They show how this trapped charge in oxide layer surrounding the piezoresistor can induce a depletion region in the semiconductor, thus increasing the device resistance. They conclude that n-type piezoresistors with the largest cross-section will be the most radiation tolerant, though there may be a tradeoff of sensitivity vs. radiation tolerance.

4.4.1.6 Suggestions for Radiation-Hardening MEMS

The difference in sensitivity of MEMS devices to radiation is due primarily to the different impact that trapped charge in dielectric layers has on different actuations schemes and geometries. MEMS operating on electrostatic principles are the most sensitive to charge accumulation in dielectric layers. In contrast, thermally and electromagnetically actuated MEMS are much more radiation tolerant. MEMS operating on piezoresitive principles, while not showing any threshold for radiation sensitivity, do not fail catastrophically until doses of several Mrad are exceeded.

Techniques that eliminate or minimize charging effects were discussed in the section on charging above, and as discussed in the RF MEMS case study in Chapter 2, include:

- Ensuring that all conductors be at well-defined potentials and not be allowed to float to avoid undesired electrostatic forces (due to charging of conductors)
- Change of dielectric material to one with lower trap density, see e.g. [114].
- Adding a charge dissipation layer on the dielectric [49].
- A geometry change to eliminate the dielectric from between moving surfaces, and from under moving surfaces.
- A geometry change to minimize the exposed area of dielectric, or replacing the dielectric films with arrays of dielectric posts [56]
- A geometry change to reduce the sensitivity to trapped charge, e.g., stiffer restoring springs.
- Electrical shielding, by covering exposed dielectric with a conductor as at well-defined potential, as in [99].

Since electrothermal and electromagnetic actuation principles are intrinsically more radiation tolerant than electrostatic operation, these actuation principles should be considered for applications where high radiation doses are expected.

4.4.2 Anodic Oxidation and Galvanic Corrosion of Silicon

4.4.2.1 Origin of Anodic Oxidation

Because of its electrical and mechanical properties as well as its relative ease of processing, poly-silicon has become the material of choice in surface-micromachined MEMS. Single-crystal silicon (often SOI) is the standard material for bulk micromachined devices. In dry ambients, such as the atmosphere found inside a package hermetically sealed in a dry and inert environment, poly-Si and single-crystal electrodes show truly impressive longevity: no signs of degradation or corrosion are observed after several months at fields close to dielectric breakdown (i.e., at fields well above those encountered during normal device operation).

Many commercial MEMS devices operate in an ultra-dry ambient in a hermetically sealed package so that the chip operates in a dry ambient even if the package is subjected to high relative humidity. If the package is non-hermetic (e.g., plastic packaging as is now most commonly the case for cost reasons), there can be leakage currents on the dielectric between neighboring electrodes, which are often unpassivated for simplicity, and to avoid dielectric charging.

If the ambient is not perfectly dry there will be several monolayers of water on any hydrophilic surface, such as the native oxide on silicon. This adsorbed water on the surface of the dielectric between electrodes provides a leakage path for current to flow. The surface current is given by [115]:

$$j_{\text{water}} \sim A \exp\left[b.RH\right] \exp\left[\frac{-E_{\text{activation}}}{k_{\text{B}}T}\right]$$

where RH is the percent relative humidity, A and b are constants, $b \sim 0.1$–0.3, $E_{\text{activation}}$: 0.4–1.1 eV, T is the temperature. Effectively no leakage occurs for humidity levels below 50%.

Anodic oxidation occurs when there is a finite surface leakage current between neighboring poly-Si electrodes on the surface of the insulator *in the presence of moisture*. The poly-Si at the anode reacts with OH^- to form SiO_2 [115].

The reaction governing anodic oxidation is:

$$H_2O \rightarrow H^+ + OH^-$$
$$Si + 2OH^- + 2h^+ \rightarrow SiO_2 + H_2 \qquad (4.12)$$
$$2H^+ + 2e^- \rightarrow H_2$$

As can be seen from (4.12), only the positively biased electrode (supplying holes, labeled h^+) is oxidized, and hence increases in volume, whereas the negatively biased electrode (supplying electrons, e^-) is unaffected.

If a poly-Si wire is allowed to fully oxidize it becomes an open circuit, and the electrode it drives becomes non-functional. Partial oxidation of the electrodes can lead to a change in the capacitance between the electrode and the electrostatically actuated part, and thus to a change in the device characteristics (e.g., mirror tilt vs.

voltage). Another reported failure mode related to anodic oxidation is delamination of poly-silicon electrodes from the dielectric [116].

4.4.2.2 Observations and Mitigation

Several groups have observed anodic oxidation in a number of MEMS geometries, including microshutter arrays [117], test structures on the SUMMiT V^{TM} process [118], and PolyMUMPS (Multi-User MEMS Process, provided by MEMSCAP) -based micromirror arrays and test structures [119] and [49].

Shea et al. report that when polysilicon test chips are operated in ambients with a RH of greater than 50%, the most positively biased unpassivated poly-Si electrodes anodically oxidize within hours or minutes, as illustrated in Fig. 4.64 for different bias conditions at 88 %RH for a 100 V bias and a 2 μm gap between the electrodes [119].

Fig. 4.64 Optical micrographs of two poly-Si electrodes on SiN$_x$, labeled A and B, with a 2 μm gap between them, showing the effect of different voltage drives at 88% RH and 23°C after 20 h of stress. *Top left*: 0 V to A&B. *Top right*: A: +100 V dc, B: grounded. Bottom left: A: -100 V dc B: grounded. *Bottom right*: A: 110 V rms, 50 Hz square wave (bipolar), B: grounded. Adapted from [119] reprinted with permission Copyright 2000 Society Of Photo Optical Instrumentation Engineers

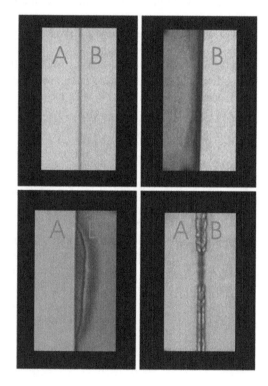

To determine the acceleration factors, Shea et al [49] subjected unsealed poly-Si test structures to both high relative humidities and high voltages. The test structures consisted of two several-hundred micron long poly-Si electrodes separated by either a 2 or a 3-μm gap. The poly-Si is the Poly0 level of the MEMSCAP MUMPS process, 500 nm thick, n$^+$ doped from a sacrificial phosphosilicate glass layer. The electrodes are electrically insulated from the substrate wafer by 600 nm of Si-rich silicon nitride. There are no moving poly-Si structures on the test chip, but the chip was released in hydrofluoric acid (HF) as are most standard surface micromachined parts (see the next section for the influence of HF on galvanic corrosion of poly-silicon).

The amount of anodic oxide that grew on the positively biased electrodes was measured after 24 h and is plotted in Fig. 4.65 (top) for four RH levels. There appears to be a threshold in relative humidity ($\sim$50%) below which anodic oxidation does not occur. No such threshold is observed for voltage. Figure 4.65 (bottom) is a plot of the total charge that flowed to an electrode over 24 h. In view of the strong correlation between the total charge flow and the measured anodic oxide height, the rate of anodic oxidation can be determined by simply measuring the surface leakage current.

It also follows from (4.12) that the rate of anodic oxidation is proportional to the leakage current between electrodes on the surface of the SiN insulator. This provides a very quick way to gauge whether anodic oxidation is occurring by simply

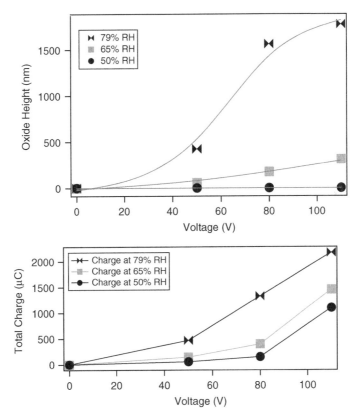

Fig. 4.65 Height of anodic oxide on anode (*top graph*) and integrated leakage current (*bottom graph*) vs. applied voltage and relative humidity, showing how the anodic oxidation is accelerated by both factors. Adapted from [49] reprinted with permission Copyright 2004 IEEE

measuring the surface leakage current. Experimentally the efficiency of this process is of the order of 2% (i.e., one SiO_2 molecule is formed for every 100 electrons that flow). The total amount of oxide grown is proportional to the total charge flow (the time integral of the surface leakage current).

Relative humidity is an accelerating factor because the higher the humidity, the more water is adsorbed on the surface, and thus the larger the surface leakage current will be (there will also be more water available to supply the OH^-) [115]. The surface leakage current increases roughly exponentially with relative humidity, and we find that, with all other conditions kept constant, the rate of anodic oxidation scales similarly.

Voltage is a strong accelerating factor because the leakage current is roughly proportional to the applied voltage. The electric field is also an accelerating factor (at a fixed leakage current). Much more oxidation is seen at sharp corners where the field is concentrated. Changing the gap between electrodes or wires from 2 to 3 μm has a large effect on the rate of anodic oxidation.

A different anodic oxidation-based failure mechanism was reported by Plass et al. at Sandia National Laboratories [120]: the progressive delamination of Polysilicon electrodes from silicon nitride layers. The authors were performing a study of dormancy-induced stiction by actuating polysilicon cantilever beams at 100 V and holding them for extended periods under different temperature and humidity conditions. The devices were made using the SUMMiT™ process.

It was observed, see Fig. 4.66, that positively-biased polysilicon actuation electrodes underneath the polysilicon cantilevers were the delaminating from the nitride, swelling and curling and fracturing the cantilever's above them. Because of the

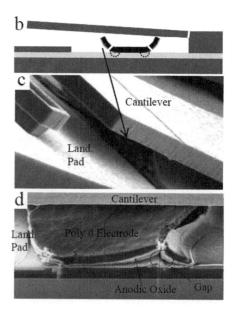

Fig. 4.66 Sandia cantilever device, on which anodic oxidation delamination of poly-silicon from silicon nitride was observed, after being held at 50% RH for 50 day at 25°C at a 100 V bias. (**b**) Schematic cross-section, (**c**) SEM image (side view) of the cantilever, (**d**) SEM view of a FIB cut through the electrode. Reference [120] reprinted with permission Copyright 2003 Society of Photo Optical Instrumentation Engineers

volume increase during oxidation and because of curling from the induced stress, the delamination mechanically interfered with the device operation well before the electrode is fully oxidized. They noted that that the delamination only starts at electrode edges directly under cantilevers, suggesting the oxidation rate also depends on the perpendicular electric field strength. In significant anodic oxidation was observed at 25% RH, but important oxidation was seen at 50% RH. Unlike Shea et al., Plass et al. see an important accelerating factor in temperature.

Hon et al. [84] report cathodic oxidation of polysilicon at high bias (>100 V) and high humidity, for the case where one of the polysilicon electrodes are in electrical contact with the substrate. They suggest that the cathodic oxidation is due to OH⁻ accumulation and a reduction in the surface potential.

Mitigating anodic oxidation requires either minimizing the operating voltages, careful choice of voltage polarity (cathodic protection), hermetic packaging or environmental control since anodic oxidation occurs extremely slowly at low humidity.

4.4.2.3 Galvanic Corrosion During Release in HF

HF is very commonly used as the final release step in processing MEMS devices because of its excellent selectivity towards silicon, and its rapid etch rate of silicon oxides. While it is known that 49% HF does not significantly affect the morphology and materials properties of single crystal silicon there have been numerous reports of HF affecting the material properties of polysilicon in particular when metallic layers are present on top of the polysilicon [29, 30, 96, 95, 121, 122].

Metal traces or coatings are often unavoidable on poly-silicon as they are required for electrical interconnections, as optical reflectors, RF antenna, or for stress control. The potential differences between the metal devices and the silicon results in galvanic corrosion. This corrosion occurs only in the HF bath and is therefore limited to the release step.

Galvanic corrosion of polysilicon is the process quite akin to the formation of porous silicon and in an HF bath, with the difference that for porous silicon the current is supplied by an external power supply, rather than by the unintended internal battery. Already in 1991 Walker et al. reported the degradation in burst strength of polysilicon membranes from varying exposures to HF [121]. A series of round robin experiments on both MCNC (now MEMSCAP) and Sandia multiuser MEMS process polysilicon found large variation in measured values of elastic properties and strength of polysilicon [123, 124]. This variation was attributed to inconsistencies in measurement techniques, variations in films growth and doping, but it is now clear that part of the differences were also due to galvanic corrosion during HF release.

The galvanic corrosion of polysilicon leads to a porous polysilicon layer, with high surface roughness (with a brown rather than shiny color). The grain boundaries in particular are etched. The Young's modulus and tensile strength of such coated polysilicon structures is greatly reduced in particular since the preferential etching at the grains leads to grooving. Figure 4.67 is a plot of tensile strength of polysilicon specimens as a function of etch time in HF [122]. The trend is

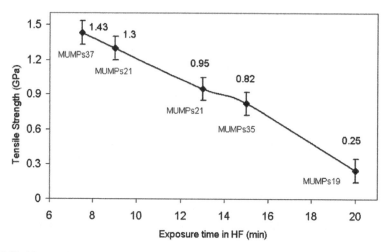

Fig. 4.67 Measured tensile strength of polysilicon test structures versus exposure time in 49% HF. Reference [122] reprinted with permission Copyright 2003 Elsevier

very clear and the effect is extremely important and can lead easily to premature device failure in view of the weakening of the polysilicon. The use of Electron Beam Scatter Detector is shown in Chapter 5 (Section 5.5.4) to allow determination of grain surface orientation, which for Polysilicon is a factor in fracture strength reduction.

While, as reported by Kahn et al., it is not clearly established whether corrosion occurs through a chemical oxidation followed by dissolution of the silicon oxide, or by direct formation of soluble $(SF6)^{2-}$ species [29], it is clearly established that the etch rate depends on the level of p- doping of the polysilicon, the length of the HF release step, and the species of metallization as well as their area coverage.

A related effect reported by Kahn et al. is the growth of very thick (up to 70 nm) native oxides on polysilicon following HF release related to the galvanic corrosion of the polysilicon. They have shown that this is linked to the p- doping, and that the effect is not seen on all polysilicon films that depends on the growth conditions.

The reduction in Young's modulus of the galvanically corroded polysilicon brings about a softening of suspensions made from such springs and associated reduction in resonant frequency. Such an effect was noted for example on the first-generation polysilicon micromirrors made by Lucent Technologies, which were made using a three-level polysilicon process on top of which are two types of metal layers, Cr-Au to serve as a highly stressed layer on the lifting arms for self-assembly, and Ti-Au to serve as a low stress optical reflector, see Fig. 4.68. The metallized chips were released in 49% HF bath for 3- 5 minutes. It was found that for springs with the expected low resistance contact (20 Ω) to the substrate, the polysilicon had the expected stiffness, and showed no mechanical creep. However, if the substrate contact, due to a processing error, was high (>1 MΩ), then galvanic corrosion took place (see Fig. 4.69), and the current assisted etching of the polysilicon led to softer

Fig. 4.68 SEM image of
polysilicon surface
micro-machined micro mirror
from Alcatel-Lucent. The
lifting arms are coated with
chrome gold, the central
0.5 mm diameter reflector is
coated with titanium gold.
Reprinted with permission of
Alcatel-Lucent USA Inc.

Fig. 4.69 Schematic
cross-section of the
micro-mirror in Fig. 4.68,
showing (*arrows*) possible
galvanic current paths leading
to etching of the silicon

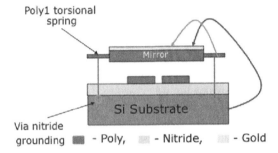

springs which were susceptible to mechanical creep (which is normally never seen
below 450°C for silicon). The corrosion problem was resolved by improving the
contact to the substrate, though other solutions were also found, such as reducing
the pH of the HF bath.

Miller et al. studied the effect of adding acids or surfactants to the HF on the
galvanic corrosion process. They found that reducing the pH of the bath by adding
HCl (4:1 aqueous HF : HCl) was very effective at reducing the galvanic corrosion
rate yielding polysilicon devices with roughness and Young's modulus comparable
to devices released without metal were released with metal but using HF vapor [96].
The effect of pH and of which reactions are favored is also discussed in [29].

Solutions to minimize galvanic corrosion include:

- Only metallize the polysilicon after the HF release (though this is often not a
  practical solution)
- Perform the release in vapor HF instead of liquid HF (this has the added benefit
  of avoiding stiction due to capillary forces when drying the device).
- Ensuring reliable substrate contacts to minimize redox potentials.

- Adding a small amount of acid (e.g., 0.1 molar HCl) to the HF bath to protect the polysilicon from corrosion.
- Minimizing the release time (which might entail changing the size and distribution of etch holes).

4.4.3 Metal Corrosion

Metal corrosion can be either chemical or electrochemical in origin. Electrochemical corrosion is more common, and is enabled by conduction through water monolayers in which contaminants have dissolved, and is driven by potentials which are either externally applied or galvanic. Corrosion is of particular concern for microelectronics and MEMS in view of the μm or sub-μm scale of metal traces: very small amounts of corrosion can lead to device failure. As many MEMS devices are sensors, some MEMS device may need to operate in harsh or corrosive environments, for instance pressure sensors for engine management or for turbine monitoring. Since, as we shall see below, corrosion can easily occur even in normal consumer application environments, we shall address primarily packaged MEMS devices that are not in direct contact with corrosive ambients.

Corrosion has three accelerating factors: (1) temperature, as temperature accelerates both diffusion and the kinetics of chemical reactions. (2) relative humidity as it provides the water surface film and, (3) concentration of trace contaminants and airborne pollutants of reactive compounds such as sulfides and chlorides that dissolve in the water surface film giving rise to high levels of corrosive ions.

Even hermetically sealed packages without getters can contain water vapor, either simply adsorbed on the surfaces, or from outgassing of water or generated from the reaction product of outgased oxygen and hydrogen. The most common packages for MEMS are plastics (e.g., overmolding on a lead frame) through which water can slowly permeate on the time scale of weeks. Most MEMS devices in use today therefore have several adsorbed monolayers of water on the surface. See also Fig. 6.21 and associated discussion.

The ionic contaminants which, when in water, lead to corrosion are of several sources [125]: (1) trace contaminants from chip processing, (2) contamination during assembly and manufacturing, (3) floods, spills, and other accidents, (4) airborne contaminants. This last category includes:

- Inorganic chlorine compounds (HCl, ClO_2, Cl_2), which produce chlorine ions in the presence of water. Sources include seawater and many household and industrial cleaning compounds.
- Sulfur compounds (H_2S, mercaptans), which can rapidly corrode copper, aluminum, and iron alloys, and whose corrosion rate is greatly accelerated by inorganic chlorine compounds. Sources include natural gas and bacteria.
- Nitrogen oxides, which can form nitric acid. The main source is the combustion of fossil fuels.
- Ozone and other strong oxidants.

From a processing perspective, SC2 cleaning (Standard Cleaning 2, also known as RCA clean) contains HCl, and vapors can deposit on surfaces, or incomplete rinsing can result in Cl⁻ at surfaces. Other cleaning baths, such as piranha (mixture of sulfuric acid and hydrogen peroxide), can generate aerosols of H_2SO_4 which become airborne in fabrication facilities. Very important is the rinsing post bath use to remove the ionics at the surface.

Corrosion is either a chemical reaction, occurring with no current flow, or an electrochemical reaction, occurring when two metals are connected electrically via an electrolyte solution; as a current flows through the electrolyte, one metal is oxidized, and the other is reduced. The driving potential can be an externally applied voltage (see anodic oxidation of silicon in an earlier section), or, for galvanic corrosion, the potential is the electric potential difference between two metals (or even for one type of metal due to local differences in morphology or composition). The main metal corrosion types are: (a) uniform, (b) galvanic (c) pitting, (d) fretting, and (e) stress corrosion cracking.

As discussed in more detail in textbooks on corrosion or on IC reliability such as [53], a commonly used method to determine safe operating potentials is the Pourbaix diagram, which displays the stable phases of metals (e.g., oxides, ions, compounds) vs. the pH of the electrolyte solution. This allows the stable phases to be identified, but does not provide information on reaction rates.

For galvanic and anodic corrosion, the corrosion speed is given by the ionic current. Conduction can occurs once only a few monolayers of water are present on a surface [126]. As shown in [127], currents of only 1 pA can cause the failure of $0.1~\mu m$ wide conductors in less than one minute from galvanic corrosion. This is particularly problematic when there are large differences in area between the two metals.

Acceleration factors are generally determined phenomenologically, with temperature and RH as the main variables. A very common test condition is 85°C at 85% RH. Table 2.3 lists several models for lifetime. The Peck model [127] can be written as:

$$MTTF = A_0(RH)^{-2.7} \exp\left[\frac{E_a}{k_b T}\right]$$

where typical activation energies E_a range from 0.7 to 0.8 eV for aluminum corrosion in the presence of chloride ions. Figure 4.70 is plot of the ratio of observed median lifetime relative to lifetime at 85°C/85%RH for aluminum in epoxy packages, showing how higher temperatures decrease lifetime.

For MEMS devices, corrosion can affect both the MEMS sensor/actuator, as well as the associated control electronics in the same package (monolithic or two-chip solutions). Approaches to minimize corrosion include:

- Passivation using barrier coatings (e.g., Parylene films, or coating with epoxy as in the "glob top" solutions), with low permeability to water, ions and gases, and excellent adhesion to avoid delamination.

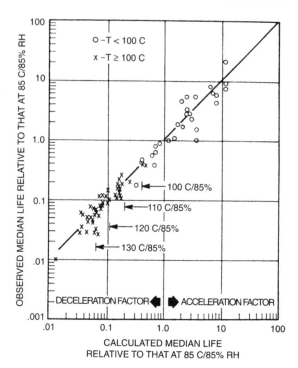

Fig. 4.70 Ratio of lifetime of aluminum metallization in epoxy packages for different conditions relative to 85°C / 85% RH conditions. Reference [127] reprinted with permission, Copyright 1986 IEEE

- Select metals with lower electric potential differences
- Select metals with better corrosion resistance
- Operate at temperatures higher than the ambient to reduce the amount of adsorbed water (e.g., TI's DMD chips are used at 10–20°C above ambient due to heating from the light source)
- Hermetic package
- Fabrication and handling changes to minimize ionic contaminants.

For MEMS pressure sensors, Bitko et al. [128] report on the reliability of Parylene (a chemical vapor deposited poly(p-xylylene) polymer) barrier coatings, which allows much lower cost packaging than using a stainless steel diaphragm and a silicone oil pressure transmission fluid, using electrochemical test methods to evaluation corrosion protection. Other encapsulants reported in Chapter 6 for Motorola pressure sensors are Fluorogels and silicone, for which corrosive media testing is reported in Section 6.5.2.

Glob top technology is popular with semiconductor overmold packaging technology for both stress relief and surface adhesion to eliminate the possibility of moisture migration at the surface to glob top interface. Chip-on-board attach methods use the glob top approach and the package itself it virtually eliminated. The glob top is often deposited over the die and wirebonds for semiconductor die, completely covering the chip and the wirebonds, providing mechanical support as well as protection from corrosion. Figure 4.71 shows that the glob top can also be deposited and cured over a capped micromachined accelerometer structure itself, and not the traces or wirebond pads.

Fig. 4.71 Glob top on accelerometer [129]. Reprinted with permission, Copyright 2008 Springer

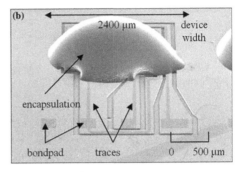

"Glob top" is a high viscosity epoxy, silicone or urethane or other material that is directly dispensed onto the component and surroundings (often including wirebonds). A related technique is the "dam and fill" in which a high viscosity filler is used to produce a ring around the component, that is then filled with a lower viscosity fill, often allowing better underfilling of parts. In both cases, the filler material protects the IC or MEMS component from the environment. The glob top materials typically have greatly reduced moisture permeability but key to high reliability is the adhesion of the glob top to the surface of the protected die, thus, the die must be clean prior to applying the glob top material. The use of glob tops is particularly important for dual-chip MEMS devices, MEMS that require a non-hermetic package such as pressure sensors and MEMs based microphones, and to enable lower cost packaging solutions in plastic packaging while protecting a MEMS capped structure such as the epoxy covered capped micromachined accelerometer in Fig. 4.71.

Glob tops can be applied to capped MEMS products for mechanical protection inside plastic overmolded packages. Hermetic packaging is much more expensive than plastic packaging. Capping of MEMS can provide a localized hermetic environment that also protects the delicate structure from microcontamination and handling (Fig. 4.72). This allows packaging in typical assembly environments which are generally too dirty for uncapped MEMS structures with their fine geometries. The ability to package a MEMS structure in a non-custom plastic overmold package will result in significant cost savings for high volume MEMS products.

MEMS packaging of the MEMtronics RF MEMS device (Chapter 2) uses a special glob top material (BCB) as an encapsulant with a SiN or Parylene sealant over it (Fig. 4.73). BCB is benzocyclobutene, also called Cyclotene (Dow Chemical). It can be spin coated for conformality and thickness control, and in this case, does not seep into the micromachined cage structure due to surface tension. As the BCB itself has finite moisture permeability, a moisture impermeable layer is sealed over

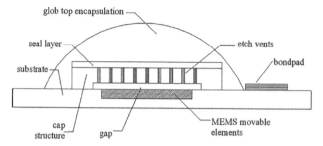

Fig. 4.72 Glob top over MEMS capped structure [129]. Reprinted with permission, Copyright 2008 Springer

Fig. 4.73 RF MEMS package using BCB as Encapsulant with sealant layer over it. Reference [130] reprinted with permission. Copyright 2005 ASME

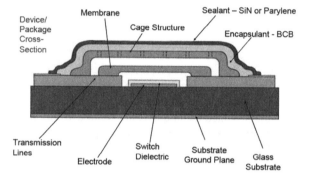

it to provide the level of hermeticity required for the MEMS device, thus generating a hermetic package by sealing the caged structure over the RF MEMS device.

Testing of glob top package solutions is performed using temperature and relative humidity, as the primary failure mechanism eliminated by this technology is corrosion. Failure of a glob top can occur due to defects in the coating or sealant structure, and poor adhesion to the surface to be protected. Coefficient of thermal expansion mismatch can also be of concern with higher modulus materials. Moisture will wick into a poorly adhered interface through capillary action and corrosion will occur in a humid field application.

Parylene, previously mentioned, is a popular coating for MEMS due to its low moisture and gas permeability and the thin film deposition technique. Parylene is chemically vapor deposited at room temperature, which allows step coverage with a thin film protective layer that can be pinhole free at layers as thin as tens of nm. Typically, microns of Parylene are deposited virtually stress free at room temperature. The biocompatibility of certain Parylenes allows MEMS structures to use this material as more than a glob top; compliant implantable sensors are being developed with Parylene.

4.5 Conclusions

This chapter covered many possible failure modes of silicon and metal MEMS devices, along with the physics of failure. The physics of failure are essential for determining the accelerated testing conditions, once the root cause has been identified (see the next chapter). This can be tricky, as one wishes to accelerate only one failure mode at a time, while many failure modes share accelerating factors. For instance temperature accelerates creep, corrosion, and dielectric breakdown. For this reason, specific test structures are often designed to investigate in isolation one failure mode. This then helps guide a redesign, or a materials change, process change, etc, leading to enhanced reliability.

The device reliability can be determined from accelerated tests, and compared to customer requests or system needs. One should note that the package is a key aspect of the MEMS reliability, as it can control the ambient the chip experiences, especially moisture, and the latter accelerates charging, corrosion, fatigue, and stiction. Packaging is very device dependent must be considered early in the design process to ensure the expected failure modes can be controlled.

References

1. Arney, S. (2001) Designing for MEMS reliability. MRS Bull 26(4), 296.
2. D. M. Tanner et al. (2000) MEMS reliability: infrastructure, test structures, experiments, and failure modes. Sandia Report SAND2000-0091, http://mems.sandia.gov/techinfo/doc/000091o.pdf
3. Gad-el-Hak, M. (ed) (2002) *The MEMS Handbook*. Boca Raton, FL: CRC Press.
4. http://www.memsnet.org/material/
5. Boyce, B.L., Grazier, J.M., Buchheit, T.E., Shaw, M.J. (2007) Strength distributions in polycrystalline silicon MEMS. J. Microelectromech. Syst. 16(2), 179.
6. Chen, K.-S., Ayon, A., Spearing, S.M. (2000) Controlling and testing the fracture strength of silicon on the mesoscale. J. Am. Ceramic Soc. 83(6), 1476–1484.
7. Sharpe, W.N., Bagdahn, J., Jackson, K., Coles, G. (2003) Tensile testing of MEMS materials—recent progress. J. Mater. Sci. 38, 4075–4079.
8. Bagdahn, J., Sharpe, W.N., Jadaan, O. (2003) Fracture strength of polysilicon at stress concentrations. J. Microelectromech. Syst. 12(3), 302–312.
9. Tanner, D.M., Walraven, J.A., Helgesen, K., Irwin, L.W., Brown, F., Smith, N.F., Masters, N. (2000) MEMS reliability in shock environments. Proc. 38th IEEE Int. Reliability Phys. Symp.
10. Rasmussen, J., Bonivert, W., Krafcik, J. (2003) High aspect ratio metal MEMS (LIGA) technologies for rugged, low-cost firetrain and control components. NDIA 47th Annual Fuze Conference, April 10. Available at: http://www.dtic.mil/ndia/2003fuze/rasmussen.pdf
11. Srikar, V.T., Senturia, S.D. (2002) The reliability of microelectromechanical systems (MEMS) in shock environments. J. Microelectromech. Syst. 11(3), 206.
12. Sundaram, S. private communication. subramanian.s88@gmail.com
13. Greek, S., Ericson, F., Johansson, S., Fürtsch, M., Rump, A. (1999) Mechanical characterization of thick polysilicon films: Young's modulus and fracture strength evaluated with microstructures. J. Micromech. Microeng. 9, 245–251.
14. Wagner, U., Muller-Fiedler, R., Bagdahn, J., Michel, B., Paul, O. (2003) Mechanical reliability of epipoly MEMS structures under shock load. TRANSDUCERS, Solid-State Sensors, Actuators and Microsystems, 12th International Conference on, vol. 1.

15. Sang, W.Y., Yazdi, N., Perkins, N.C., Najafi, K. (2005) Novel integrated shock protection for MEMS. Solid-State Sensors, Actuators and Microsystems. Digest of Technical Papers. TRANSDUCERS '05. The 13th International Conference on, vol.1.
16. Liu, H., Bhushan, B. (2004) Nanotribological characterization of digital micromirror devices using an atomic force microscope. Ultramicroscopy 100, 391–412.
17. Hartzell, A., Woodilla, D. (1999) Reliability methodology for prediction of microma-chined accelerometer stiction. 37th International Reliability Physics Symposium (IRPS), San Diego, California, 202
18. Swiler, T.P., Krishnamoorthy, U., Clews, P.J., Baker, M.S., Tanner, D.M. (2008) Challenges of designing and processing extreme low-G microelectromechanical system (MEMS) accelerometers, Proc. SPIE 6884, 68840O, DOI:10.1117/12.77 .
19. http://www.colibrys.com/files/pdf/products/DS%20HS8000%20D%2030S.HS8X%20F.03.09.pdf
20. Stauffer, J.-M., Dutoit, B., Arbab, B. (2006) Standard MEMS sensor technologies for harsh environment. IET Digest 2006, (11367), 91–96, DOI:10.1049/ic:20060450 .
21. Stauffer, J.-M. (2006) Standard MEMS capacitive accelerometers for harsh environ-ment. Paper CANEUS2006-11070, Proceedings of CANEUS2006, August 27–September 1, Toulouse, France
22. Habibi, S., Cooper,S.J., Stauffer, J.-M., Dutoit, B. (2008) Gun hard inertial measurement unit based on MEMS capacitive accelerometer and rate sensor. IEEE/ION Position Location and Navigation System (PLANS) Conference.
23. Ghose, K., Shea, H.R. (2009) Fabrication and testing of a MEMS based Earth sensor. Transducers 2009, Denver, CO, USA, June 21–25, paper M3P.077.
24. Suhir, E. (1997) Is the maximum acceleration an adequate criterion of the dynamic strength of a structural element in an electronic product? IEEE Trans. Components Packaging Manuf. Technol.—Part A, 20(4), December 1997.
25. Harris C.M. (ed) (2002) Harris' Shock and Vibration Handbook, 5th edn. New York: McGraw-Hill.
26. Young, W.C. Roark's Formulas for Stress And Strain, 6th edn. New York: McGraw Hill.
27. Muhlstein, C. L., Brown, S. B., Ritchie, R. O. (2001) High-cycle fatigue of single-crystal silicon thin films. J. Microelectromech. Syst. 10, 593.
28. Gasparyan, A., Shea, H., Arney, S., Aksyuk, V., Simon, M.E., Pardo, F., Chan, H.B., Kim, J., Gates, J., Kraus, J.S., Goyal, S., Carr, D., Kleiman, R. Drift-Free, 1000G Mechanical Shock Tolerant Single-Crystal Silicon Two-Axis MEMS Tilting Mirrors in a 1000×1000-Port Optical Crossconnect, Post deadline paper PD36-1, Optical Fiber Communication Conference and Exhibit 2003, OFC 2003, March 2003 Atlanta, GA. DOI: 10.1109/OFC.2003.1248617
29. Kahn, H., Deeb, C., Chasiotis, I., Heuer, A.H. (2005) Anodic oxidation during MEMS processing of silicon and polysilicon: native oxides can be thicker than you think. J. Microelectromech. Syst. 14, 914–923.
30. Miller, D., Gall, K., Stoldt, C. (2005) Galvanic corrosion of miniaturized polysilicon struc-tures: morphological, electrical, and mechanical effects. Electrochem. Solid-State Lett. 8, G223–G226.
31. Modlinski, R., Ratchev, P., Witvrouw, A., Puers, R., DeWolf, I. (2005) Creep-resistant aluminum alloys for use in MEMS. J. Micromech. Microeng. 15, S165–S170, doi:10.1088/0960-1317/15/7/023
32. Douglass, M.R. (1998) Lifetime estimates and unique failure mechanisms of the digi-tal micromirror device (DMD). 36th Annual International Reliability Physics Symposium, Reno, Nevada.
33. Modlinski, R., Witvrouw, A., Ratchev, P., Puers, R., den Toonder, J.M.J., De Wolf, I. (2004) Creep characterization of Al alloy thin films for use in MEMS applications. Microelectron. Eng. 76, 272–278.
34. Ritchie, R.O., Dauskardt, R.H. (1991) J. Ceram. Soc. Jpn. 99, 1047–1062.
35. Ritchie, R.O. (1999) Mechanisms of fatigue-crack propagation in ductile and brittle solids. Int. J. Fracture 100, 55–83.

36. Van Arsdell, W.W., Brown, S.B. (1999) Subcritical crack growth in silicon MEMS. J. Microelectromech. Syst. 8, 319.
37. Muhlstein, C.L., Stach, E.A., Ritchie, R.O. (2002) Mechanism of fatigue in micron-scale films of polycrystalline silicon for microelectromechanical systems. Appl. Phys. Lett. 80, 1532.
38. Muhlstein, C.L., Stach, E.A., Ritchie, R.O. (2002) A reaction-layer mechanism for the delayed failure of micron-scale polycrystalline silicon structural films subjected to high-cycle fatigue loading. Acta Mater. 50, 3579.
39. Kahn, H., Chen, L., Ballerini, R., Heuer, A.H. (2006) Acta Mater. 54, 667
40. Kahn, H., Ballerini, R., Bellante, J.J., Heuer, A.H. (2002) Fatigue failure in polysilicon not due to simple stress corrosion cracking. Science 298, 1215.
41. Alsem, D.H. et al. (2007) Very high-cycle fatigue failure in micron-scale polycrystalline silicon films: effects of environment and surface oxide thickness. J. Appl. Phys. 101, 013515.
42. Douglass, M.R. (2003) DMD reliability: a MEMS success story. In R. Ramesham, D. Tanner (eds) Proceedings of the Reliability, Testing, and Characterization of MEMS/ MOMES II, SPIE, Bellingham, WA, Vol. 4980, 1–11.
43. Yang, Y., Allameh, S., Lou, J., Imasogie, B., Boyce, B.L., Soboyejo, W.O. (2007) Fatigue of LIGA Ni micro-electro-mechanical system thin films. Metallurgical Mater. Trans. A 38A, 2340.
44. Cho, H.S., Hemker, K.J., Lian, K., Goettert, J., Dirras, G. (2003) Sens. Actuat. A: Phys. 103(1–2), 59–63.
45. Son, D. et al. (2005) Tensile properties and fatigue crack growth in LIGA nickel MEMS structures. Mater. Sci. Eng. A 406, 274–278.
46. Allameh, M., Lou, J., Kavishe, F., Buchheit, T.E., Soboyejo, W.O. (2004) Mater. Sci. Eng. A 371, 256–66.
47. Reid, J.R., Webster, R.T. (2002) Measurements of charging in capacitive microelectromechanical switches. Electron. Lett. 38(24), 1544–1545.
48. De Groot, W.A., Webster, J.R., Felnhofer, D., Gusev, E.P. (2009) Review of device and reliability physics of dielectrics in electrostatically driven MEMS devices. IEEE Trans. Dev. Mater. Reliability 9(2), art. no. 4813209, 190–202.
49. Shea, H.R., Gasparyan, A., Chan, H.B., Arney, S., Frahm, R.E., López, D., Jin, S., McConnell, R.P. (2004) Effects of electrical leakage currents on MEMS reliability and performance. IEEE Trans Dev. Mater. Reliability 4(2), 198–207.
50. Crank, J. (1997) The mathematics of diffusion, 2nd edn. Oxford: Clarendon Press, 47–51.
51. Lewis, T.J. (1978) In D.T. Clark and W.J. Feast (eds) The Movement of Electrical Charge Along Polymer Surfaces in Polymer Surfaces. New York: John Wiley & Sons.
52. Ehmke, J., Goldsmith, C., Yao, Z., Eshelman, S. (2002) Method and apparatus for switching high frequency signals. Raytheon Co., United States patent 6,391,675; May 21.
53. Ohring, M. (1998) Reliability and Failure of Electronic Materials and Devices. New York: Academic Press, 310–325 and references therein.
54. Rebeiz, G.M., Muldavin, J.B. (2001) RF MEMS switches and switch circuits. in IEEE Microwave Magazine 2(4), 59–71.
55. Rebeiz, G.M. (2003) RF MEMS: Theory, Design and Technology. New York: John Wiley and Sons.
56. Goldsmith, C.L., Forehand, D., Scarbrough, D., Peng, Z., Palego, C., Hwang, J.C.M., Clevenger, J. (2008) Understanding and improving longevity in RF MEMS capacitive switches. Proc. Int. Soc. Opt. Eng. 6884(03), Feb 2008.
57. Peng, Z., Palego, C., Hwang, J.C.M., Moody, C., Malczewski, A., Pillans, B., Forehand, D., Goldsmith, C. (2009) Effect of packaging on dielectric charging in RF MEMS capacitive switches. IEEE Int. Microwave Symp. Dig. 1637–1640, June 2009.
58. Wibbeler, J., Pfeifer, G., Hietschold, M. (1998) Sens. Actuators A 71, 74–80.
59. Van Spengen, W.M., Puers, R., Mertens, R., De Wolf, I. (2004) A comprehensive model to predict the charging and reliability of capacitive RF MEMS switches. J. Micromechan. Microeng. 14(4), 514–521.

60. Goldsmith, C.L., Ehmke, J., Malczewski, A., Pillans, B., Eshelman, S., Yao, Z., Brank. J., Eberly, M. (2001) Lifetime characterization of capacitive RF MEMS switches. IEEE MTT-S Int. Microwave Symp. Digest 3, 227–230.

61. Peng, Z., Palego, C., Hwang, J.C.M., Forehand, D., Goldsmith, C., Moody, C., Malczewski, A., Pillans, B., Daigler, R., Papapolymerou, J. (2009) Impact of humidity on dielectric charging in RF MEMS capacitive switches. IEEE Microwave Wireless Comp. Lett. vol. 1.

62. Schönhuber, M.J. (1969) Breakdown of gases below paschen minimum: basic design data of high-voltage equipment. IEEE Trans. Power Apparatus Syst. vol. PAS-88, 100, Feb 1969

63. Dhariwal, R.S., Torres, J.M., Desmulliez, M.P.Y. (2000) Electric field breakdown at micrometre separations in air and nitrogen at atmospheric pressure. IEE Proc. Sci. Meas. Technol. 147(5), 261–265.

64. Torres, J.-M., Dhariwal, R.S. (1999) Electric field breakdown at micrometre separations. Nanotechnology 10, 102–107.

65. Slade, P.G., Taylor, E.D. (2002) Electrical breakdown in atmospheric air between closely spaced (0.2 μm–40 μm) electrical contacts. IEEE Trans. Comp. Packaging Technol. 25 (3), 390–396.

66. Wallash, A., Levit, L. (2003) Electrical breakdown and ESD phenomena for devices with nanometer-to-micron gaps. Proc. of SPIE 4980, 87–96.

67. Chen, C.-H., Yeh, J.A., Wang, P.-J. (2006) Electrical breakdown phenomena for devices with micron separations. J. Micromech. Microeng. 16, 1366–1373.

68. Strong, F.W., Skinner, J.L., Tien, N.C. (2008) Electrical discharge across micrometer-scale gaps for planar MEMS structures in air at atmospheric pressure. J. Micromech. Microeng. 18, 075025.

69. Paschen, F. (1889) Über die zum Funkenübergang in Luft, Wasserstoff und Kohlensäure bei verschiedenen Drucken erforderliche Potentialdifferenz. Annalen der Physik 273(5), 69–96.

70. Townsend, J. (1915) *Electricity in Gases*. New York: Oxford University Press.

71. Braithwaite, N.St.J. (2000) Introduction to gas discharges. Plasma Sources Sci. Technol. 9, 517–527.

72. Osmokrovic, P., Vujisic, M., Stankovic, K., Vasic, A., Loncar, B. (2007) Mechanism of electrical breakdown of gases for pressures from 10-9 to 1 bar and inter-electrode gaps from 0.1 to 0.5 mm. Plasma Sources Sci. Technol. 16, 643–655.

73. Torres, J.-M., Dhariwal, R.S. (1999) Electric field breakdown at micrometre separations in air and vacuum. Microsyst. Technol. 6(1), November, 6–10, DOI 10.1007/s005420050166 .

74. Carazzetti, P., Shea, H.R. (2009) Electrical breakdown at low pressure for planar MEMS devices with 10 to 500 micrometer gaps. J. Micro/Nanolithography, MEMS, and MOEMS 8(3), 031305.

75. Habermehl, S., Apodaca, R.T., Kaplar, R.J. (2009) On dielectric breakdown in silicon-rich silicon nitride thin films. Appl. Phys. Lett. 94, 012905.

76. Amerasekera, A., Duvvury, C. (2002) *ESD in Silicon Integrated Circuits*, 2nd edn. New York: John Wiley and Sons.

77. Walraven, J.A., Soden, J.M., Tanner, D.M., Tangyunyong, P., Cole Jr., E.I., Anderson, R.E., Irwin, L.W. (2000) Electrostatic discharge/electrical overstress susceptibility in MEMS: A new failure mode. Proceedings.

78. Ruan, J., Nolhier, N., Papaioannou, G.J., Trémouilles, D., Puyal, V., Villeneuve, C., Idda, T., Coccetti, F., Plana, R. Accelerated lifetime test of RF-MEMS switches under ESD stress. Microelectron. Reliability 49(9–11), 125.

79. Tazzoli, A., Peretti, V., Meneghesso, G. (2007) Electrostatic discharge and cycling effects on ohmic and capacitive RF-MEMS switches. IEEE Trans. Dev. Mater. Reliability 7(3), 429–436.

80. Sangameswaran, S., Coster, J.D., Linten, D., Scholz, M., Thijs, S., Haspeslagh, L., Witvrouw, A., Hoof, C.V., Groeseneken, G., Wolf, I.D. (2008) ESD reliability issues in microelectromechanical systems (MEMS): A case study on micromirrors. Electric.

81. Krumbein, S.J. (1987) Metallic electromigration phenomena 33rd meeting of the IEEE Holm Conference on Electrical Contacts, Published by AMP Inc, 1989. http://www.tycoelectronics.com/documentation/whitepapers/pdf/p313-89.pdf

82. Black, J.R. (1969) IEEE Trans. Electron Dev. ED-16, 338.

83. Courbat, J., Briand, D., de Rooij, N.F. (2008) Sens. Actuators A 142, 284–291.

84. Hon, M., DelRio, F.W., White, J.T., Kendig, M., Carraro, C., Maboudian, R. (2008) Cathodic corrosion of polycrystalline silicon MEMS. Sens. Actuators A: Phys. 145–146, July–August 2008, 323–329, DOI: 10.1016/j.sna.20

85. Stark, B. (1999) MEMS reliability assurance guidelines for space applications. JPL Publication 99-1. http://trs-new.jpl.nasa.gov/dspace/bitstream/2014/18901/1/99-9001.pdf

86. Shea, H. (2006) Reliability of MEMS for space applications. Proc. of SPIE Reliability, Packaging, Testing, and Characterization of MEMS/MOEMS V, Vol. 6111, 61110A. DOI:10.1117/12.651008 .

87. "Handbook of radiation effects", by A. Holmes-Siedle and L. Adams, Oxford University press, 2nd edition, 2002

88. European Cooperation for Space Standardization, document ESCC Basic Specification 22900 for Total Dose Steady-State Irradiation Test Method, available at: https://escies.org/ReadArticle?docId=229

89. European Cooperation for Space Standardization, document ESCC Basic Specification 25100 for Single Event Effects Test Method and Guidelines. Available at: https://escies.org/ReadArticle?docId=229

90. SPENVIS, the Space Environment Information System, http://www.spenvis.oma.be/

91. Shea, H. (2009) Radiation sensitivity of microelectromechanical system devices. J. Micro/Nanolith. MEMS MOEMS 8(3), 031303, Jul–Sep 2009.

92. European Space Agency Procedures Standards and Specifications, document ESA PSS-01-609 (May 1993) Radiation Design Handbook, available at: https://escies.org/ReadArticle?docId=263

93. European Cooperation for Space Standardization, document ECSS-E-ST-10-04C Space environment, available at: http://www.ecss.nl/forums/ecss/dispatch.cgi/standards/showFile/100700/d20081115082809/No/ECSS-E-ST-10-04C (15 November2008).pdf

94. Beasley, M. et al. (2004) MEMS thermal switch for spacecraft thermal control. Proc. SPIE 5344(98), DOI:10.1117/12.530906 .

95. Pierron, O.N., Macdonald, D.D., Muhlstein, C.L. (2005) Galvanic effects in Si-based microelectromechanical systems: thick oxide formation and its implications for fatigue reliability. Appl. Phys. Lett. 86, 211919.

96. Miller, D.C., Hughes, W.L., Wang, Z.L., Gall, K., Stoldt, C.R. (2006) Galvanic corrosion: a microsystems device integrity and reliability concern. Proc. SPIE 6111, 611105, DOI:10.1117/12.644932 .

97. Coumar, O., Poirot, P., Gaillard, R., Miller, F., Buard, N., Marchand, L. Total dose effects and SEE screening on MEMS COTS accelerometers. Radiation Effects Data Workshop, 2004 IEEE 22 July 2004, 125–129.

98. Lee, C.I., Johnston, A.H., Tang, W.C., Barnes, C.E. (1996) Total dose effects on micromechanical systems (MEMS): accelerometers. IEEE Trans. Nucl. Sci. 43, 3127–3132.

99. Knudson, A.R., Buchner, S., McDonald, P., Stapor, W.J., Campbell, A.B., Grabowski, K.S., Knies, D.L. (1996) The effects of radiation on MEMS accelerometers. IEEE Trans. Nucl. Sci. 43, 3122–3126.

100. Edmonds, L.D., Swift, G.M., Lee, C.I. (1998) Radiation response of a MEMS accelerometers: an electrostatic force. IEEE Trans. Nucl. Sci. 45, 2779–2788.

101. Schanwald, L.P. et al. (1998) Radiation effects on surface micromachined comb drives and microengines. IEEE Trans. Nucl. Sci. 45(6), 2789–2798.

102. Holbert, K.E., Nessel, J.A., McCready, S.S., Heger, A.S., Harlow, T.H. (2003) Response of piezoresistive MEMS accelerometers and pressure transducers to high gamma dose. Nuclear Sci. IEEE Trans. 50(6), Part 1, Dec. 2003, 18.

103. McCready, S.S. et al. (2002) Piezoresistive micromechanical transducer operation in a pulsed neutron and gamma ray environment. IEEE Radiation Effects Data Workshop, 181–186.
104. Marinaro, D. et al. (2008) Proton radiation effects on MEMS silicon strain gauges. IEEE Trans. Nuclear Sci. 55(3), 1714.
105. Quadri, G., Nicot, J.M., Guibaud, G., Gilard, O. (2005) Optomechanical microswitch behavior in a space radiation environment. IEEE Trans. Nuclear Sci. 52(5), 1795.
106. Miyahira, T.F. et al. (2003) Total dose degradation of MEMS optical mirrors. IEEE Trans. Nuclear Sci. 50(6), Part 1, Dec. 2003, 1860–1866.
107. McClure, S., Edmonds, L., Mihailovich, R., Johnston, A., Alonzo, P., DeNatale, J., Lehman, J., Yui, C. (2002) Radiation effects in microelectromechanical systems (MEMS): RF relays. IEEE Trans. Nucl. Sci. 49, 3197–3202, Dec. 2002.
108. Tazzoli, A., Cellere, G., Peretti, V., Paccagnella, A., Meneghesso, G. (2009) Radiation sensitivity of OHMIC RF-MEMS switches for spatial applications. Proc. IEEE Int. Conference Micro Electro Mechanical Syst. (MEMS).
109. Buchner, S. et al. (2007) Response of a MEMS microshutter operating at 60 K to ionizing radiation. IEEE Trans. Nuclear Sci. 54(6), 2463.
110. Caffey, J.R., Kladitis, P.E. (2004) The effects of ionizing radiation on microelectromechanical systems (MEMS) actuators: electrostatic, electrothermal, and bimorph. Micro Electro Mech. Syst. 17th IEEE Int. Conference MEMS.
111. Son, C., Ziaie, B. (2008) An implantable wireless microdosimeter for radiation oncology. Proc. IEEE Conference Micro Electro Mech. Syst. (MEMS 2008), 256.
112. Schwartz, R.N. et al. (2000) Gamma-ray radiation effects on RF MEMS switches. Proc. 2000 IEEE Microelectron. Reliability Qualification Workshop, Oct. 2000, IV.6.
113. Zhu, S.-Y. et al. (2001) Total dose radiation effects of pressure sensors fabricated on Unibond-SOI materials. Nucl. Sci. Tech. 12, 209–214.
114. Lamhamdi, M. et al. (2006) Characterization of dielectric-charging effects in PECVD nitrides for use in RF MEMS capacitive switches. Proc of 7th International conference on RF MEMS and RF microsystems (MEMSWAVE) 2006.
115. Comizzoli, R.B. (1991) Surface Conductance on Insulators in the Presence of Water Vapor, in Materials Developments in Microelectronic Packaging: Performance and Reliability. Proceedings of the Fourth Electronic Materials and Processing Congress, 311
116. Lewerenz, H.J. (1992) Anodic oxides on silicon. Electrochimica Acta 37, 847–864.
117. Perregaux, G., Gonseth, S., Debergh, P., Thiebaud, J.-P., Vuilliomenet, H. "Arrays of addressable high-speed optical microshutters" MEMS 2001. The 14th IEEE International Conference on MEMS, Jan 2001, 232–235.
118. Plass, R.A., Walraven, J.A., Tanner, D.M., Sexton, F.W. Anodic oxidation-induced delamination of the SUMMiT poly 0 to Silicon Nitride Interface. In R. Ramesham, D.M. Tanner (eds) Reliability, Testing, and Characterization of MEMS/MOEMS II; Proc. SPIE V.
119. Shea, H.R., White, C., Gasparyan, A., Comizzoli, R.B., Abusch-Magder, D., Arney, S. (2000) Anodic oxidation and reliability of poly-Si MEMS electrodes at high voltages and in high relative humidity, in MEMS Reliability for Critical Applications, R.A. Lawton Proc. SPIE, 4180, 117–122. DOI: 10.1117/12.395700
120. Plass, R., Walraven, J., Tanner, D., Sexton, F. (2003) Anodic oxidation-induced delamination of the SUMMiT poly 0 to silicon nitride interface. In Proc. SPIE, 4980, 81–86.
121. Walker, J.A., Gabriel, K.J., Mehregany, M. (1990) Mechanical integrity of polysilicon films exposed to hydrofluoric acid solutions. In: Proceedings from IEEE MEMS, Napa Valley, CA, February 11–14, 56–60.
122. Chasiotis, I., Knauss, W.G. (2003) The mechanical strength of polysilicon films: part 1. The influence of fabrication governed surface conditions. J. Mech. Phys. Solids 51, 1533–1550.
123. Sharpe, W.N., Brown, J.S., Johnson, G.C., Knauss W.G. (1998) Round-robin tests of modulus and strength of polysilicon. Materials Research Society Proceedings, Vol. 518, San Francisco, CA, pp. 57–65.

124. LaVan, D.A., Tsuchiya, T., Coles, G., Knauss, W.G., Chasiotis, I., Read, D. (2001) Cross comparison of direct strength testing techniques on polysilicon films. In Muhlstein, C., Brown, S.B. (eds) *Mechanical Properties of Structural Films*, ASTM STP.

125. Sinclair, J.D. (1988) Corrosion of electronics, the role of ionic substances. J. Electrochem. Soc. March 1988, p. 89C.

126. Yan, B.D., Meilink, S.L., Warren, G.W., Wynblatt, P. (1986) Proc. Electron. Components Conf., 36, 95.

127. Peck, D.S. (1986) Comprehensive model for humidity testing correlation. Annual Proceedings – Reliability Physics (Symposium), 44–50.

128. Bitko, G., Monk, D.J., Maudie, T., Stanerson, D., Wertz, J., Matkin, J., Petrovic, S. Analytical techniques for examining reliability and failure mechanisms of barrier-coated encapsulated silicon pressure sensors exposed.

129. Hamzah, A.A., Husaini, Y., Majlis, B.Y., Ahmad, I. (2008) Selection of high strength encapsulant for MEMS devices undergoing high-pressure packaging. Microsyst.Technol. 14(6), 766.

130. Forehand, D.I., Goldsmith, C.L. (2005) Wafer Level Micropackaging for RF MEMS Switches. 2005 ASME InterPACK '05 Tech Conf, San Francisco, CA, July 2005.

131. Rebeiz, G.M. (2003) RF MEMS switches: status of the technology. Proceedings of Transducers 2003, The 12th International Conference on Solid State Sensors, Actuators and Microsystems, Boston, June 8–12, 1726.

132. Koons, H.C., Mazur, J.E., Selesnick, R.S., Blake, J.B., Fennell, J.F., Roeder, J.L., Anderson, P.C. (1670) The impact of the space environment on space systems. Aerospace Corp. report no. TR-99 (1670)-1, 20 July 1999.

133. George, T. (2003) Overview of MEMS/NEMS technology development for space applications at NASA/JPL. Proceedings of SPIE, Proc. SPIE Int. Soc. Opt. Eng. 5116, 136.

134. Eberl, C. et al. (2006) Ultra high-cycle fatigue in pure Al thin films and line structures. Mater. Sci. Eng. A, 421(1–2), 15 April 2006, 68–76.

135. Greywall, D. et al. (2003) Crystalline silicon tilting mirrors for optical cross-connect switches. IEEE/ASME Journal of Microelectromechanical Systems 12(5), 708

136. Arney, A., Gasparyan, A., Shea, H., SPIE Short course 434, Designing MEMS for reliability, presented at SemiCon West 2003, San Francisco, CA, USA

Chapter 5
Root Cause and Failure Analysis

5.1 Introduction

This chapter will cover strategies for identifying root cause and corrective action of reliability field failures. The MEMS reliability program must include strategies for identifying potential failure modes, failure mechanisms, risk areas in design and process, and containment strategies. Containment of the failure is crucial to achieving a low field failure rate while the root cause is determined and the proper corrective action is developed, checked for effectiveness, and then finally implemented into production.

Failure analysis results from yield loss parts, burn-in failures, accelerated test failures, and reliability field failures are part of the reliability program (Chapter 2). In this chapter, use of the proper failure analysis analytical techniques will be covered. Combining failure analysis results with mechanical data, electrical test data, and production process in-line metrics can provide key information about failure mechanisms. Analysis of yield loss in the production line is important to understanding early life failure mechanisms as well as steady state failure. Proper use of this information can identify trends in production populations, and determine how to eliminate the failures from occurring in both production (Chapter 3) and the field (Chapter 4).

A recommended starting place is use of the Failure Modes and Effects Analysis (FMEA) methodology. For emerging technologies, the use of proven methodologies to identify potential failure modes and mechanisms is a structured approach. The FMEA has sections to be evaluated and filled in by the team. All failure modes that can possibly occur should be included, using the cause and effect approach.

- "How the failure occurs" is the potential failure mode, also called physics of failure.
- The effect is the "consequence on the system" under study; questions such as "what parameter(s) are possibly changed due to this failure mode" are asked.
- Safeguards are put into place as "prevention measures" during the design and manufacturing phases. For example, design for 'reliability is performed in the early phases of the product development (Chapter 7).

A.L. Hartzell et al., *MEMS Reliability*, MEMS Reference Shelf,
DOI 10.1007/978-1-4419-6018-4_5, © Springer Science+Business Media, LLC 2011

- "Actions" are risk evaluations on the occurrence of the physics of failure. If the risk is high, actions also are performed as a way to mitigate or eliminate failure. Actions include reliability testing, inspections, and data gathering and analysis.

Each potential reliability risk area is given three quantitative values (Sect. 5.2.1). Potential reliability failure modes are listed, the FMEA is filled in, and quantitatively triaged. This allows the highest risk areas to be focused on first with cost/benefit in mind.

5.2 FMEA, Failure Mode and Effects Analysis

FMEAs are traditionally used for manufacturing process and design. The PFMEA (P=process) and the DFMEA (D=design) are already required by many MEMS customers in high reliability applications such as automotive or space missions. A standard that defines the FMEA is the SAE J 1739 Reference Manual [1], yet there are many more. The FMEA approach is recommended here to be adopted for MEMS reliability and will be called the RFMEA (R=reliability).

A flow chart used to generate a RFMEA is shown in Fig. 5.1 [2]. Generation of a RFMEA is bounded by identification of the goals of the project. Next, a team of engineers and scientists are chosen to generate the FMEA within the defined scope.

Many inputs (failure modes, safeguards, etc.) for the RFMEA will be found within the PFMEA and DFMEA documents. Whenever a detection solution (see definition of detection in the next paragraph) results in a reliability test, this must be added to the reliability FMEA. Do not think that your only inputs come from the PFMEA and DFMEA documents; there will be unique failure mechanisms only identified in the reliability FMEA.

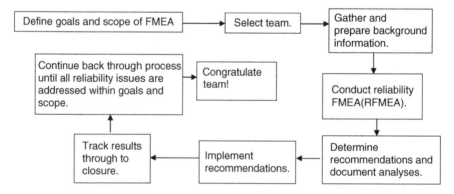

Fig. 5.1 An example flow chart for generation of reliability FMEA

5.2.1 RPN (Risk Priority Number) Levels

The format of the reliability FMEA (RFMEA) can match the format of process and design FMEA, however, any format is appropriate as long as it works for the team. Suggested herein is a rigorous format that includes a RPN (risk priority number) quantitative value for each identified potential failure mode on the RFMEA. The RPN value is obtained by multiplying three values.

- The first value is severity. For each failure mode, a severity (S) value is assigned. This value is from 1 to 10, where 1 is typically defined as a failure that is not noticeable and does not affect the product performance, while a severity of 10 is the most extreme failure and can result in death.
- The second value is occurrence, which is assigned via the "effect" of the failure mode. The likelihood of an occurrence (O) value is how often the effect is predicted to be observed. Again, a 1–10 scale is given to occurrence where a 1 is very unlikely and a 10 is inevitable.
- The third value is detection (D). This is a function of effectiveness of the prevention or mitigation measures in place. Methods for detection include inspection and test within the manufacturing facility, or reliability testing prior to shipment to the customer. However, if the methods cannot bring out a failure mode (or screen it) and the customer will experience it, then the detection level is 10. The failure mode cannot be detected! If there is a 100% certainty that controls are in place which will result in catching the failure prior to shipment to the customer and the customer will not experience this particular failure mode, then the detection level is 1.

5.2.2 RFMEA Example

An example of a RFMEA is presented in Table 5.1, without the RPN values for ease of presentation. It is understood that this is the partial result of a final RFMEA. An RPN example follows in Table 5.2. An optical MEMS mirror switch and package is the subject of this RFMEA. Included are an item number for tracking, a description of the component that has a probability of failure, the failure mode or "cause" (how the component will fail), the consequences on the system (the "effect") and the safeguards for detection. These safeguards are prevention or mitigation measures put into place with the goal to reduce the detection value to a target of 1. Also recommended is a column for actions. In an RFMEA, the actions are typically reliability-based but are not restricted to this.

Table 5.2 includes an example from Table 5.1 (Item No. 3) with RPN values included, so the reader understands how to use this valuable quantitative tool. Item

Table 5.1 MEMS reliability FMEA with highest RPN values listed [2] (reprinted with permission. Copyright 2007 SPIE)

Item No.	Description	Failure mode	Effect	Safeguards	Actions
Number for component	*Name of component*	*How it fails*	*Consequences on system*	*Prevention or mitigation measures in place*	*Actions needed to eliminate, reduce, or mitigate failure risk level*
(1) Mirror component A.	A mirror of specific diameter and thickness, and suspension spring dimensions.	Mechanical shock and stresses experienced during lifetime creates fracture in actuator springs/hinges, or allows stiction to occur.	Resonant frequency changes and/or fracture occurs, or in the case of stiction, restoring force is not large enough to allow mirror to spring back from surface.	Perform design for reliability analysis to assure (a) actuator springs/hinges are designed to withstand shock and mechanical cycles, (b) restoring force is adequate.	Perform reliability testing for shock and cyclic fatigue. Monitor changes in resonant frequency and inspect for fracture. Inspect and test for stiction.
(2) Package	A packaging scheme, with optical fiber attach.	Tolerance deltas over life change relative positions of mirrors, and could result in hermeticity loss.	Insertion loss is increased and is higher than specification limits, and in the case of hermeticity loss, moisture related failures such as corrosion can occur.	Perform design for reliability analysis prior to choosing packaging piece-parts and processes.	Perform reliability testing and compare mirror shifts to predicted tolerance changes. Accelerated testing for hermeticity loss.

Table 5.1   (continued)

Item No.	Description	Failure mode	Effect	Safeguards	Actions
Number for component	*Name of component*	*How it fails*	*Consequences on system*	*Prevention or mitigation measures in place*	*Actions needed to eliminate, reduce, or mitigate failure risk level*
(3) Mirror component B.	A mirror of specific diameter and thickness with known starting curvature and known reflective coating.	Curvature exceeds specification limits due to coefficient of thermal expansion mismatch which results in system mirror curvature match changes	Insertion loss is higher than specification limits due to fiber coupling efficiency degradation.	Perform design for reliability on mirror to evaluate curvature matching.	Perform thermal testing and compare radius of curvature change to coupling loss predictions.
(4) Mirror component C.	A mirror with reflective coating consisting of, at minimum, an adhesion layer and a reflective layer.	Diffusion of mirror substrate and/or adhesion layer through reflective layer with potential for oxidation at surface.	Reflectivity is significantly reduced, insertion loss is high and outside of specified limits.	Choose materials that do not diffuse through reflective layer in subsequent thermal exposure or through lifetime.	Perform accelerated testing, measure reflectivity before and after tests. Perform materials analysis on layered films.

Table 5.2 One item is evaluated for RPN values in a RFMEA, example

Item No.	Description	Failure mode	(S)	Effect	(O)	Safeguards	(D)	RPN
Number for component	*Name of component*	*How it fails*		*Consequences on system*		*Prevention or mitigation measures in place*		
3) Mirror component.	A mirror of specific diameter and thickness with known starting curvature and known reflective coating.	Curvature exceeds specification limits due to coefficient of thermal expansion mismatch which results in mirror curvature match problems.	9	Insertion loss is higher than spec limits due to fiber coupling efficiency loss.	2	Perform design for reliability on mirror to optimize curvature matching in design.	3	54
						Test for curvature with an interferometer tool or equivalent.	2	36
						Perform accelerated thermal testing and compare radius of curvature change to curvature matching modeling.	2	36

No. 3 is the mirror component with a known geometry and initial curvature. The failure mode is a due to mirror curvature match change during thermal exposure, resulting in fiber coupling efficiency reduction. As the coupling efficiency is a contributor to insertion loss, a critical parameter in the optical switch system, the severity (S) is 9 in the case of failure – the product will fail if the curvature exceeds the specification limits, but death will not occur. The occurrence (O) is lower, a 2. There is a low risk that this will be experienced by the customer. This is because the detection (D) is effective: safeguards have a high probability of eliminating this failure mode in the design phase (D = 3), in test (D = 2), and through accelerated testing (D = 2).

5.3 Case Study of RFMEA Failure Mode

The previous example assumes the reliability engineer and RFMEA team perform their work properly to reduce any field failure risk. The three actions in Table 5.2 are performed to gather data on the risk identified in the RFMEA. In this case study, the fiber coupling loss will be modeled with focus on the mirror curvature matching in the optical switch system to understand the effect that mirror curvature changes will have on performance in the field.

5.3.1 RFMEA Safeguard: Design for Reliability, Mirror Curvature Matching

Next, predictive modeling for fiber coupling efficiency on the MEMS mirror based system is performed. (It is recommended that readers review reference 2 which contains a comprehensive modeling analysis of a 2×2 optical MEMS switch, as seen in Figs. 5.2 and 5.3.) An unfolded view of the switch in one state is shown in Fig. 5.3.

In this optical MEMS switch state, X2 and X3 are MEMS optical mirrors, while X1 and X4 are lenses coupled to input and output fibers. L is the length of the optical path of the two lenses, FL is the focal length of the lenses coupled to fibers, and the beam waists at various locations in the system are represented.

Mirror curvature of both mirrors versus the fiber coupling efficiency is modeled and results are presented in Fig. 5.4 [2]. The radius of the beam is changed in this design as a function of mirror curvature, as the beam in Gaussian in profile, and the curved mirror would act as a lens. A flat mirror would simply redirect the light, which is optimal.

- Case A occurs when mirrors X2 and X3 have the same curvature.
- Case B is the case when the X3 mirror is flat versus a curved X2.
- Case C is when the X3 mirror has opposite sign of curvature as X2.

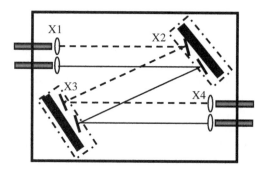

Fig. 5.2 Top view of an optical 2 × 2 MEMS switch. Two input fibers enter from the *left*, two output fibers exit on the *right*. The four micro-mirrors allow light to be switched from any input to any output. Two of the four possible light paths are shown as *one solid* and *one dashed line*. Reference [2] reprinted with permission. Copyright 2007 SPIE

Fig. 5.3 Unfolded path through one switching state of device. X2 and X3 are the MEMS mirrors, X1 and X4 are the collimating lenses. Reference [2] reprinted with permission. Copyright 2007 SPIE

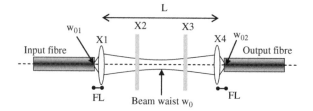

Both cases B and C in Fig. 5.4 provide some compensation to the curvature of X2, yet the worst case condition of Case A is likely typical, as the mirror design and processing conditions are assumed to be the same. Note that curvature in this figure is dependent on mirror dimensions and temperature variations, which were held constant during modeling. We assign D = 3 in Table 5.2 for the modeling as a relatively low risk because this modeling shows high coupling efficiency for a range of achievable mirror curvature matches.

For background on how the mirror design and processing conditions can affect initial curvature and curvature over operating temperature, equations are presented that relate mirror curvature to geometric, thermal and mechanical parameters. First, the Stoney equation can be used to calculate the stress in the thin film of the mirror. The basic Stoney equation [3] for full wafer bow of a thin film on a thick substrate with zero stress is shown here.

$$\sigma_f = \frac{E_s t_s^2}{6(1 - \nu_s)t_f R} \ \ (Pa) \tag{5.1}$$

In Eq. (5.1), σ_f is the film stress, R is the radius of curvature (here in μm obtained through wafer bow measurement using a technique such as a KLA-Tencor FLX tool), E_s is the Young's modulus of the substrate (typically silicon in units of MPa), t_s and t_f are the substrate and film thicknesses, respectively (in μm), and ν_s is the

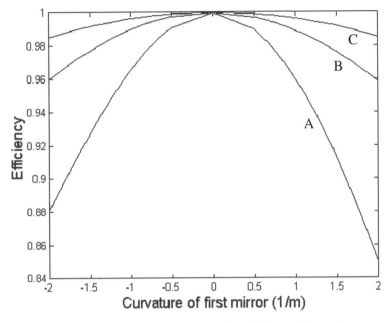

Fig. 5.4 Two-mirror optical switch curvature vs. coupling efficiency when varying curvature of each mirror independently. Reference [2] reprinted with permission. Copyright 2007 SPIE

Poisson's ratio of the substrate (unitless). Radius of curvature can also be measured with an interferometer tool, often applied to single MEMS mirrors (Sect. 5.5.2). The Stoney equation is valid for single micromachined mirrors, as long as the mirror thickness is much larger than the metal coating thickness, and as long as the suspension does not induce too much deformation of the mirror.

Total stress in a thin film is typically a summation of the thermal, intrinsic and externally applied stresses (Eq. (5.2)):

$$\sigma_{tot(f)} = \sigma_{th} + \sigma_{int} + \sigma_{ext} \tag{5.2}$$

The thermal stress is defined in Eq. (5.3). Intrinsic stress is a complex function of the structure of the deposited thin film which includes materials defects, grain boundaries, deposition rate, etc., and is typically very small compared to the thermal stress. Externally applied stresses are assumed to be negligible in this example.

The following expression equates the thermally induced stress with the change in temperature ΔT through use of the Young's modulus of the reflective film E_f (often gold, aluminum or silver), Poisson's ratio of the reflective thin film, and the coefficients of thermal expansion of the reflective film and substrate, respectively, α_f and α_s. Thermal stress is present as the deposition is performed at elevated temperature where the thin film is strain-free. Upon cooling post deposition, a thermal stress is present that is proportional to the difference of the thermal coefficients of expansion of the thin film and substrate. ΔT is thus generally the difference between the deposition temperature and normal operating temperature, or between the lowest temperature at which the metal can rapidly creep (and hence will be stress-free) and normal operating temperature.

$$\sigma_{th} = \frac{E_f}{1 - \nu_f}(\alpha_s - \alpha_f)\Delta T \tag{5.3}$$

Equations (5.1) and (5.3) are equated to evaluate R, the radius of curvature of the mirror component as a function of temperature. Use of this set of equations will allow a prediction of curvature in the MEMS mirror in various cases while varying thin film thickness, thin film materials set, substrate thickness, and temperature range. In order for this approach to have realistic results, the thickness of the thin film (reflective film) must be very small compared to the substrate thickness. The thin film must also have stable stress properties that are non-anisotropic, uniform, have a bow-free substrate, and be unconstrained externally (Fig. 5.5).

Fig. 5.5 Wafer curvature illustration. Reference [4] Copyright 2002 Defence Technical Information Center, Air Force Institute of Technology

5.3.2 RFMEA Safeguard: Test for Curvature

As cited in Section 5.3.1, curvature testing can be performed with a blanket thin film deposited on a substrate, typically a silicon wafer. The Tencor FLX or equivalent tooling is commonly used in semiconductor fabs and controlled environment factories that deposit thin films for monitoring of thin film stress for various applications. The tool uses a laser that scans along one axis of the wafer, measuring radius of curvature and wafer bow.

For mirror curvature accuracy, however, measurement of the MEMS mirror itself is preferred. Mirror substrate surfaces, thicknesses, and process variations can result in different results versus measurement of a thin film on a full sized single crystal silicon wafer, or polysilicon mirror surface. Interferometry is recommended as the technique of choice for curvature measurements. An image of a Boston Micromachines Corporation gold-coated segmented mirror is shown in this section. Single mirror segments are fantastic indicators of post thin film deposition stress, as the polysilicon substrate is quite thin (<10 microns) for this MEMS mirror. The radius of curvature as well as the RMS surface roughness can be determined with the measurement tool. Figure 5.6 is an interferometric image of one segmented mirror with post reflective thin film deposition radius of curvature (RoC or R) of 9.64

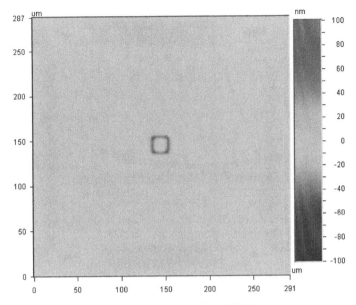

Fig. 5.6 Surface Topography of Boston Micromachines MEMS mirror segment measured with interferometric tool. Courtesy P. Bierden, Boston Micromachines Corporation

meters and Rq (RMS roughness <6 nm). The RMS Roughness Rq is defined as

$$R_q = \sqrt{\frac{1}{N}\sum_i^N Z_i^2} \qquad (5.4)$$

Where N is the number of sample points taken during the measurement and Z_i is the distance from the measured point to the mean plane of the sample. K, known as the curvature, is the inverse of the radius of curvature (see Fig. 5.7). In this case, K = 0.1037 m^{-1}. Figure 5.4 shows >99% coupling efficiency for this curvature in the model of the optical switch system.

The curvature requirements for MEMS micro-mirrors are a function of the use application. As the operating environment typically varies in temperature, stages (set-ups) are available that can vary the temperature of the MEMS mirror during interferometric measurement so that radius of curvature measurements can be performed over the entire thermal operating range of the system. Comparison to the results of the Stoney equation approach detailed in this section can be performed. The D = 2 value in Table 5.2 is given due to the existence of excellent toolsets to measure curvature and the ability to accurately measure curvature over temperature, and to compare to modeling results.

One possible root cause in this MEMS mirror example is excessive curvature change due to high thermal stress. Corrective action can include

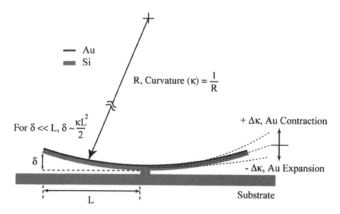

Fig. 5.7 Image depicting radius of curvature for a thin gold reflective film on a silicon substrate. Reference [5] reprinted with permission. Copyright Elsevier 2004

- Lowering deposition temperature.
- Reduce reflective coating (thin film) thickness if reflectivity is not deleteriously affected.
- Coating both sides of the mirror (this can lead to long term curvature if the creep rates do not match on both sides of the mirror due to different deposition conditions).
- Using a metal that has a high creep rate to allow rapid reduction in film stress at room temperature.

The first two solutions will decrease the thermal stress and allow the mirror to stay within specification over operating temperature. The second two solutions concern creep and long-term evolution of stress in the metal films, discussed in the next section. Process engineering experts will work with reliability engineers to perfect the corrective action to reduce thermal stress.

5.3.3 RFMEA Safeguard: Perform Accelerated Thermal Testing and Compare Radius of Curvature Change to Predictions

Due to coefficient of thermal expansion differences with reflective coatings and mirror substrates and possible long term plastic deformation effects such as creep (Chapter 4), accelerated temperature testing is recommended to assure that continuous operation at elevated temperatures will not permanently deform the mirror and create out of specification curvature conditions. The activation energies of some thermally activated failure mechanisms are in the literature, and some are not (Chapter 2 discussed acceleration of thermally activated creep). For your materials set, if the activation energy is not known, an Arrhenius experiment can be performed to determine the activation energy that results in curvature, and then lifetime predictions can be made for various operation conditions. As this is an accepted model to develop thermally-activated acceleration factors, a low risk value of $D = 2$ is given in the RFMEA.

The Arrhenius method takes some time and many samples. If the product under study and the company has these resources, this will allow for the most accurate lifetime predictions. A minimum of four temperatures are chosen and samples are put into categories by temperature. (It is critical that the same failure mechanism be accelerated at these temperatures, and it must be the same failure mechanism observed in the product use environment.) At various read-out points performed at increasing intervals (such as time zero, time = 50 h, time = 100 h, time = 200 h, time = 500 h, time = 1000 h), each set of parts is taken from the oven, allowed to cool and stabilize, and measured on the interferometer for radius of curvature. Sample sizes are a function of part availability, yet for a new device, it is important to perform this well and to assure sufficient sample sizes to determine an accurate activation energy, with minimal influence of infant mortality.

The changes in radius of curvature are plotted as % change vs. time, for each temperature. Next, the time at which 50% (or another appropriate metric) of the samples are out of specification (have failed) is plotted for each temperature on a curve as in Fig. 5.8. The Arrhenius equation is of the form:

$$rate \propto \exp\left[\frac{-E_a}{kT}\right]$$

The data is plotted as the natural log of the time to 50% failure on the y axis, and $1/T$ (°K) is plotted on the x axis. The slope of the line produced is $-E_a/k$ where k is Boltzmann's constant. Multiplication of the slope with Boltzmann's constant will give the activation energy, Ea, of the curvature mechanism (the failure mechanism in this example). Typically, this is a creep-related phenomenon for gold coated samples, which would be the *root cause* of failure. Note if the data is not linear in this plot, the mechanism is not logarithmic and the Arrhenius method cannot be used as the acceleration model.

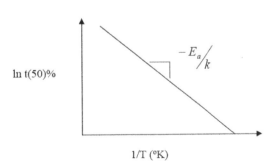

Fig. 5.8 Arrhenius method
used to determine activation
energy

The change in curvature over lifetime for thermally activated mechanisms can be accurately predicted by the acceleration factor model in Chapter 2, repeated below in Eq. (5.5).

$$A_F = e^{Ea/k\left(\frac{1}{T_{use}} - \frac{1}{T_{accel}}\right)}$$

(5.5)

The curvature change over the product life is predicted over the entire operating temperature range, compared against the specification, calculated for lifetime failure rate, and put into the RFMEA. Safety-critical applications have failure rate limits of 1 part per million or lower. If the failure rate of curvature change at the end of life is more than can be permitted for the fiber coupling efficiency (in this case), process and/or design changes must be studied and implemented to reduce the lifetime failure rate. Alternatively, creep-related failure can be reduced by materials choices or process enhancements such as post deposition annealing. Identification of the proper root cause and corrective action requires careful evaluation.

5.3.4 Implementation of RFMEA Learning into Production

The concept of Think-Do-Check-Act is used for evaluation. In the Think-Do-Check-Act method, the engineering team determines the root cause of the failure (failure mode), develops a corrective action (design or process changes), and then carefully implements the corrective action into production. Full implementation into production without first checking the correction action is never recommended unless the situation is dire. The "Check" portion of the method is performed as follows: three lots of material are run through the production line with the new fix. The material is fully tested and analyzed to assure that the corrective action did not introduce any unknown and/or unwanted deleterious effects to the final product [6]. It is important to remember that some mechanisms are not able to be "tested out". Thus, additional reliability testing such as burn-in and/or accelerated testing on the corrective action material is performed.

Curvature matching over life and operating temperature for a MEMS optical switch was highlighted by an RFMEA and detailed in this chapter. However, for any failure mechanism in which corrective action is implemented into production, at minimum, limited reliability testing is highly recommended prior to full corrective action implementation. Why?

A. Burn-in can catch early life failures that are typically defect-related (infant mortality in Section 2.3); burn-in makes sure these parts do not make it to the customer and fail early in lifetime. Designing proper burn-in methodologies for your particular MEMS device requires answers to the following questions:

   a. What stress testing will create failure quickly for your particular failure mechanisms but won't result in harming the good parts? (see Table 2.1 and Tables 6.1 and 6.2).

b. Are you putting too much lifetime on the parts prior to shipment?

c. Do you need separate or combined electrical and mechanical type burn-in methods (or another stress test for your MEMS application)?

d. Reliability testing is covered in some qualification specifications, such as Mil-Std-883; these can be used as background information when developing burn-in testing. Some useful methods are cited next.

    i. Method 1005, Steady State Life, is often performed for 24–48 h at 125°C as a burn-in method.

    ii. Method 1006, Intermittent Life, has power on-off cycling at temperature during test. Again, the time period of testing would be decreased to simulate the infant mortality portion of the bathtub curve.

    iii. Method 1007, Agree Life, is a combination of thermal stress, power on-off cycling, and vibration testing.

B. Accelerated testing (reliability testing – Section 2.2) will generate data on the useful life failure rate of the MEMS device as well as predict wear-out. Predictive methodologies exist which can compare the field failure rates with the accelerated test failure predictions. A mature MEMS device with the proper reliability program will find these two failure rates are very similar.

5.4 Failure Analysis as a Tool for Root Cause

Critical to determine root cause is performance of failure analysis. Failure analysis should be performed on yield loss parts, burn-in failures, accelerated test failures (Chapter 2), and field failures. Performing failure analysis is different as a function

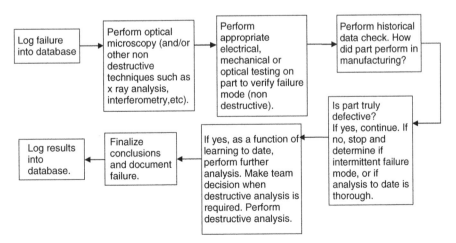

Fig. 5.9 Flow chart for failure analysis

of the part type as well as the failure mode. A typical flow chart for failure analysis
is shown in Fig. 5.9.

5.5 Analytical Methods for Failure Analysis

Analytical methods for failure analysis or characterization of MEMS parts are
too numerous to present, thus, a few key techniques will be covered and examples presented. Dynamic analysis (Laser Doppler Vibrometry), structural techniques
(Interferometry, SEM, EBSD, FIB, TEM, AFM) and chemical/compositional techniques (EDS, Auger, TOFSIMS, XPS, FTIR) are presented.

5.5.1 Laser Doppler Vibrometry (LDV)

The Laser Doppler vibrometer is a very powerful technique that provides dynamic
information of MEMS devices. In LDV, the MEMS device is driven with a periodic AC signal, resulting in surface movement while an interferometric technique is
detects the path-length shift of the reflected laser beam. For the Polytec^{TM} system,
Mach-Zehnder interferometry is used. The transient response of an overdamped
optical MEMS device is shown in Fig. 5.10.

A Fourier transform can provide acceleration, velocity, or displacement amplitude vs. frequency in the FFT mode (Fig. 5.11), allowing rapid identification of the
resonance frequencies.

The resonance frequencies of different modes and determination of which modes
are excited can be detected and recorded. The square two-axis mirror in Fig. 5.12
produces the resonance frequencies in Fig. 5.13. The frequency was swept under a
sine-wave excitation of the mirror to produce the dynamic response. Figure 5.14 is

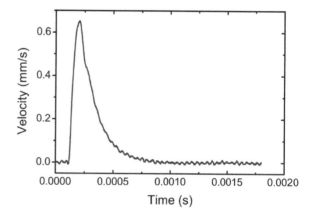

Fig. 5.10 LDV in transient
mode, showing the response
to a step in drive voltage for
an optical MEMS device (for
an overdamped mode).
Courtesy, S. Sundaram, EPFL

Fig. 5.11 Displacement vs. frequency signal in FFT domain LDV for an optical MEMS device. Courtesy, S. Sundaram, EPFL

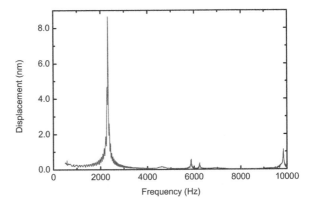

Fig. 5.12 Two axis mirror that produces resonances in Fig. 5.13. Reference [7] reprinted with permission. Copyright 2000 SPIE

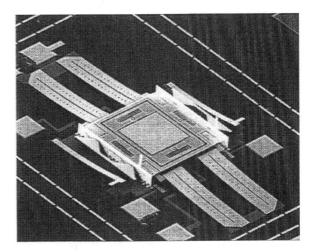

a long silicon cantilever actuated in the first out-of-plane mode, showing the maximum deflection position of the 14.47 kHz mode for a given drive amplitude. It is overlaid on an optical microscope image of the un-actuated device. The z-scale is given in the figure and is exaggerated compared to the in-plane scale to allow easy visualization. The cantilever is 1 mm long and 80 μm thick.

The laser vibrometer also provides animation of recorded data: an adjustable mesh can cover a single device or entire array, and the laser beam of the laser vibrometer scans set-points within the mesh and records data from different devices within a few seconds. LDV tools exist that can measure out of plane single point vibration, out of plane differential vibration, in plane vibration, rotational vibration, 3D vibration, and both 2D and 3D surface vibration mapping. LDV with stroboscopic video microscopy is covered in Chapter 6. As a failure analysis tool, this technique can be used to measure resonance frequency reduction during cyclic fatigue (Chapter 4).

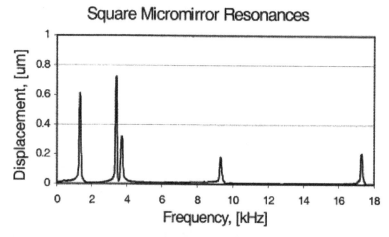

Fig. 5.13 Resonances measured using LDV for the two-axis square mirror. Reference [7] reprinted with permission. Copyright 2000 SPIE

Fig. 5.14 Silicon cantilever at maximum deflection in first out-of-plane mode using LDV. Courtesy J. Gomes, EPFL

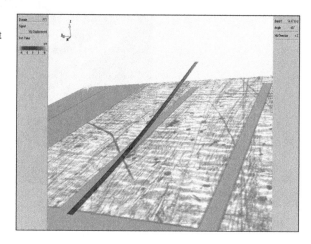

5.5.2 Interferometry

Surface profiling interference microscopes are typically used to image surfaces in 3D with very high precision, up to 100X magnification (Fig. 5.15). An illustration of an interference microscope using a Michelson configuration can be seen in Fig. 5.16. In this configuration, the objective contains a beam splitter to divide the illumination into a reference and sample beam; upon reflection from a reference mirror and the sample surface, the two beams recombine and interfere. The resulting interferogram is made up dark and light "fringes", and contains information on surface topography of the sample. A CCD camera registers this interferogram at various reference arm phase shifts; which are then processed by mapping algorithms to produce the

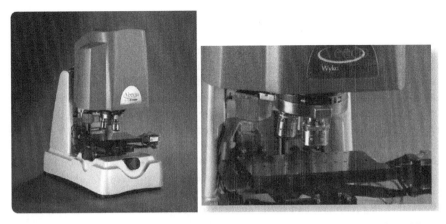

Fig. 5.15 Veeco Wyko NT9100. Reference [8] reprinted with permission. Image courtesy of Veeco Instruments Inc.

Fig. 5.16 PSI representation of Veeco Interferometer NT1100 [9]. Reprinted with permission. Image courtesy of Veeco Instruments Inc.

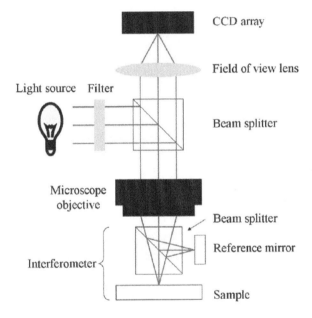

sample topography information. Figure 5.6 of this chapter is an image generated by an interferometer. In this figure, PSI (phase shifting interferometry) mode was utilized; monochromatic light is used in PSI and has sub-nanometer resolution for very smooth surfaces. For the flat segmented mirror in Fig. 5.6, this is an extremely accurate technique. However, if steps > $\lambda/4$ are present and/or the surface is rough, VSI (vertical scanning interferometry) is often used. VSI uses white light and generates fringes at the best focus position. This technique has nanometer vertical resolution.

5.5.3 Scanning Electron Microscopy (SEM)

Scanning electron microscopy (SEM) is a fundamental tool for MEMS analysis. Feature sizes of MEMS devices are on the order of a micron, thus, the wavelength of light is too large for detailed delineation of structure and surface during imaging. The electron beam of the SEM is directed via a magnetic lens to a sample; upon interaction of the beam with the sample, secondary electrons are emitted or backscattered and detected by a secondary electron detector or backscattered detector, respectively. The electrons collected from the sample provide a surface image that is topographical in nature (secondary electron image) with atomic number and orientation information (backscattered electrons). Sources for electron emission have gotten more complex over the years, yet the primary sources are tungsten filament, LaB_6 filament, and field emission source [10]. The field emission source is often preferred in MEMS imaging due to the high brightness versus the tungsten and LaB_6 sources. For Bio-MEMS and structures that are mostly polymer based, special near-atmospheric tungsten filament SEM instruments are preferred as they do not require high vacuum pump-down and therefore have less electron charging at the sample surface.

The field emission SEM (FE-SEM) image in Fig. 5.17 is an image of polysilicon post HF immersion. The various grains are observed nicely, as crystalline grains have different orientations and lattice spacings, and therefore different secondary electron emission efficiencies resulting in varying grey scales. Precipitates have been etched at grain boundaries, creating voids. This image is taken at 1.99 KV; a low accelerating voltage was used to reduce charging. The sample did not need to have a conductive coating applied. Typically samples are coated with an extremely thin layer of carbon or gold or another precious metal to eliminate charging if higher accelerating voltages are used or if the sample is an insulator. This moderate magnification of 20,310X has a 200 nm scale bar so the user can determine feature sizes.

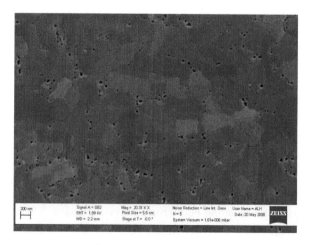

Fig. 5.17 Polysilicon post HF immersion

The depth of field of the scanning electron microscope allow for imaging of the pits where the oxygen rich precipitates were preferentially etched through HF (hydrofluoric acid) immersion. These pits form primarily at the grain boundaries. Variations in processing that generate, for example, oxygen precipitates that are subsequently etched in HF can reduce fracture strength on polysilicon [11].

5.5.4 Electron Beam Scatter Detector (EBSD)

To gain crystallographic quantitative information of polycrystalline, a detector is added to the scanning electron microscope, which is commonly referred to as EBSD (electron beam scatter detector). The electron beam contacts a tilted sample surface, electrons are diffracted and subsequently hit the fluorescent screen of the detector. The diffraction pattern gives orientation, grain boundary, and defect information. Scanning of the electron beam across the sample will result in a map of a polycrystalline sample, resulting in quantitative information about the microstructure of the sample.

The sample in Fig. 5.18 is polysilicon mapped with EBSD. Polysilicon can be degraded mechanically during release with hydrofluoric acid (HF). The example with scanning electron microscopy was for a mechanism that decreases polysilicon fracture strength due to HF etch and surface pitting. Here, a different mechanism is covered called galvanic corrosion. Galvanic corrosion is covered in depth in Section 4.4.2. In this system with polysilicon, gold and HF, gold is the cathode, polysilicon is the anode, and HF is the electrolyte.

Polysilicon surface grain orientation is a factor in fracture strength reduction through galvanic corrosion during HF release while in contact with gold metal (such as gold bond pads on the wafer). Other factors in this degradation mechanism, in addition to the surface grain orientation, are HF release time and concentration of the HF electrolyte. Analysis of the surface layer morphology has shown some polysilicon orientations are more susceptible to the galvanic corrosion degradation than others. EBSD was performed on various samples of polysilicon that were not exposed to HF. Samples were then exposed to HF, fracture strength testing was performed, and a difference was seen in galvanic corrosion resistance. Figure 5.18 shows the pre-HF EBSD results [12]. Figure 5.18(a) is the crystallographic texture map in which many of the surface grains lie in a {110} orientation parallel to the top surface of the sample. Other random grain orientations are observed for grains but these grains are not lying parallel to the top surface (The triangular scale in (a) is next to the EBSD image.). Figure 5.18(b) is an inverse pole-figure plot taken along the axis of deposition. The scale for this plot is to its right. This shows a ratio of 3:1 grains in the {110} direction versus random orientation, again primarily at the surface. This is considered a weak crystallographic texture. A stronger <110> orientation in these maps is thought to result in less electrochemical degradation of the polysilicon due to galvanic corrosion [12, 13].

Fig. 5.18 Crystallographic texture map (**a**) and corresponding inverse pole-figure plot (**b**) for non-corroded polysilicon, along the axis of deposition (with scales). Reference [12] reprinted with permission, Copyright 2008 American Institute of Physics

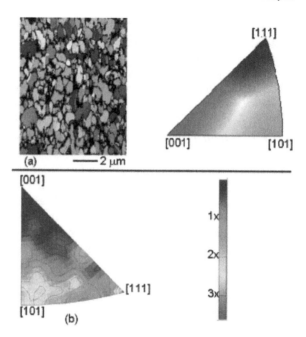

5.5.5 *Transmission Electron Microscopy (TEM)*

Transmission electron microscopy uses an electron beam like SEM, yet that electron beam is available at much higher accelerating voltages, and the sample must be thinned so that the image is collected after the beam moves through the sample. TEM produces images at very high magnifications, and be performed in diffraction mode which can provide crystallographic information for crystalline material, as well as for individual grains in polycrystalline materials. This powerful technique requires careful sample preparation, which is typically performed with the focused ion beam technique (Section 5.5.6).

In the following example, stress state of polysilicon is correlated to TEM grain imaging. Figure 5.19 illustrates the crystal structure of polysilicon in TEM images, while plotting resistivity versus stress in the film. The films are two micron undoped polysilicon with subsequent phosphorus implant and 4 hour 950°C anneal. As can be seen by the TEM images, increased deposition temperature results in smaller grain size, increased resistivity, and lower tensile stress. The 560°C deposition is amorphous with high tensile stress and low resistivity post anneal. At 580°C deposition the film is mixed phase, but the stress is less tensile in the post anneal state. The 590°C deposited film is fully polycrystalline with small grains and a very low stress post anneal. Resistivity is a good measurement due to correlation with the stress and grain structure [14]. This technique is easy to measure as the small grains limit the resistance. Low stress polysilicon films are typically preferred in MEMS structures.

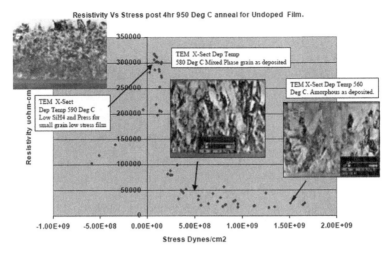

Fig. 5.19 TEM images correlated with resistivity versus stress for 2 micron polysilicon film with phosphorus implant and 4 hour, 950°C anneal. Reference [14] reprinted with permission. Copyright 2001 Vacuum and Technology Coating 2

5.5.6 Focused Ion Beam (FIB)

Focused Ion Beam sectioning is a fantastic method for investigation of MEMS failures. The dual beam system is FIB with SEM in one tool. Figure 5.20 illustrates this system for the LEO 1500 Cross Beam system [15]. This allows FIB sputtering, deposition, and ion imaging, with secondary electron imaging in one tool. There are many dual beam instruments available today, including excellent tools from Carl Zeiss.

The FIB typically uses a gallium ion beam to sputter the sample for cross section, and has the ability to deposit material (typically tungsten) to samples as well. Figure 5.21 is an illustration of the gas injection system of the FIB. The beam starts

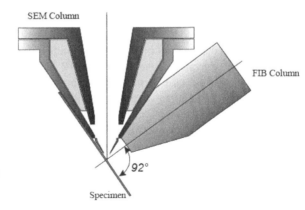

Fig. 5.20 LEO 1500 Cross Beam FIB SEM system. Reference [15] reprinted with permission. Copyright 2003 SPIE

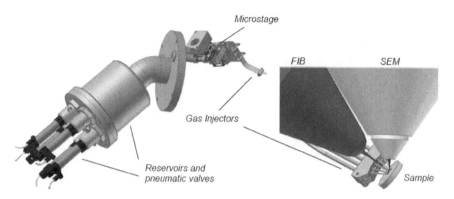

Fig. 5.21 Gas injection system for FIB dual beam system. Reference [15] reprinted with permission. Copyright 2003 SPIE

Fig. 5.22 TEM sample during final FIB polish. Reference [15] reprinted with permission. Copyright 2003 SPIE

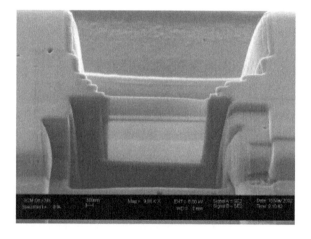

in coarse sputtering mode, and gradually reduces sputter rate to a final polish phase. A TEM section during final polish is shown in Fig. 5.22.

Due to the very precise nature of the FIB beam cuts, cross sectional analysis is a very popular use of FIB tools. With the invention of the dual beam or cross beam tools, SEM imaging in the tool allows high resolution imaging post FIB sectioning. The FIB sections in Fig. 5.23 are Sandia microengines operated for 2 million and 379 million cycles, respectively, prior to sectioning to check for debris or other anomalies [16]. (These samples were purposely removed from the stress test for destructive analysis.) The hub and pin joint regions were clear of debris, and these regions are most susceptible to wear in this design. These samples were first

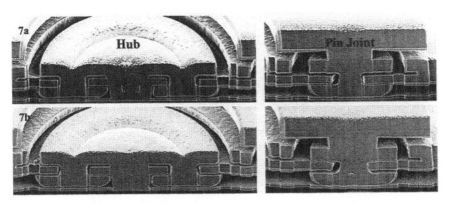

Fig. 5.23 Hub and Pin Joint FIB sections post cycling did not show evidence of wear or debris. Reference [16] reprinted with permission. Copyright 2000 SPIE

FIB coated with tungsten to protect the surface during FIB sputtering. The Sandia microengine is covered in Chapters 2, 4 and 6.

5.5.7 Atomic Force Microscopy (AFM)

Atomic force microscopy is a powerful and critical analytical tool for MEMS. It provides single atomic step height and sub-angstrom RMS surface roughness measurements. It can provide more fine surface detail than the interferometer. A fine cantilevered tip on the end of a probe scans the surface of the MEMS structure, while a feedback system maintains a constant force between the probe and the surface, at a constant deflection, termed "contact mode" which is also known as static mode. This is an imaging mode. Dynamic mode AFM is performed by oscillating the tip at close to the resonance frequency. During dynamic mode AFM, the oscillation changes at the surface are compared to a reference oscillation to provide surface information. Non-contact mode AFM operates the tip above the resonance frequency, at a low amplitude oscillation. Soft samples are better imaged in non-contact mode. More information on AFM techniques can be found with any commercial AFM manufacturer.

In Fig. 5.24, the DMD exploded view in the Chapter 2 Case Study has accompanying AFM images of each layer. In this study, surface height and friction force images were amongst many surface maps obtained with the AFM. Figure 5.25 is of the hinge array, where contact is made on the DMD. The term "finger print" is where mechanical wear has taken place, while the hinge with "no finger print" has no mechanical wear. The "finger print" image shows a smear-like feature in both the surface map and friction force image, and the surface map has a higher peak to valley range than the "no finger-print" image. This is attributed to wear of the SAM (self-assembled monolayer) coating on the hard aluminum surface [17].

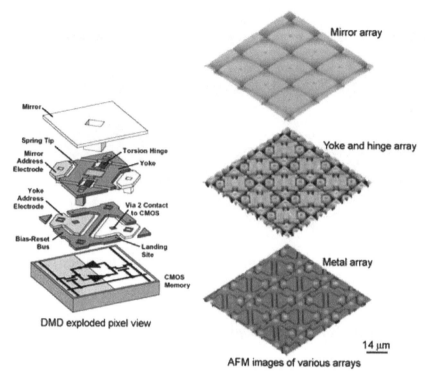

Fig. 5.24 Exploded view of DMD with AFM images of the various arrays. Reference [17] reprinted with permission. Copyright 2004 SPIE

5.5.8 Energy Dispersive X-ray Analysis (EDS, EDX, EDAX)

The high energy electron beam in electron microscopy interacts with atoms at the surface, knocking away electrons in valence shells close to the nucleus and creating vacancies. In energy dispersive x-ray analysis, commonly referred to as EDS, EDX or EDAX, a spectrum of energy peaks is generated as higher energy electrons will move to lower energy valence shells and fill these vacancies. Upon filling a lower energy shell vacancy, energy in the form of an x-ray is emitted and detected with an EDS detector that is positioned at the proper take off angle inside the electron microscope sample chamber. A spectrum of energy peaks is generated and as elements have characteristic peaks, elemental information can be gained from the sample.

Alpha and beta represent the electron that fills the vacancies. If the electron is from one valence shell higher, then the x-ray generated is called alpha. If the electron filling the vacancy comes from two valence shells away, it is called beta. In the case of various subshells (L and M valences), a subscript is added to identify the subshell that donated the electron to fill the inner vacancy. The nomenclature of these peaks is K, L or M (valence shell that lost the electron and has a vacancy) followed by α, β, α_1, α_2, for example (Fig. 5.26).

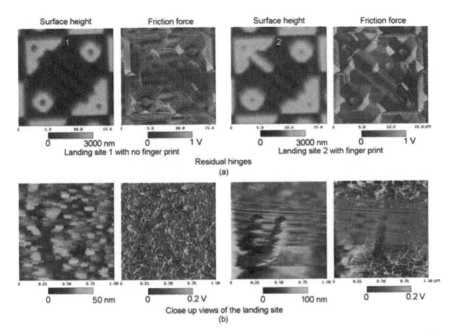

Fig. 5.25 Surface height and friction force maps generated by AFM for the hinge of the DMD product. Wear and increased surface roughness is observed on the hinge on the right in the images, termed "fingerprint". Reference [17] reprinted with permission. Copyright 2004 SPIE

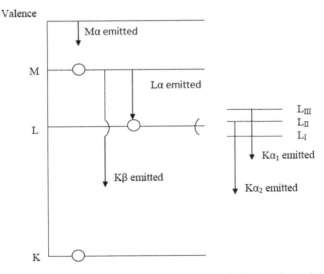

Fig. 5.26 Depiction of electrons filling lower energy valence shell vacancies and their nomenclature for EDS [18]. Reprinted with kind permission of Springer Science and Business Media, Copyright 1984

EDS is not a particularly surface sensitive technique as the interaction depth of a 20 keV accelerating voltage can be on the order of one micron. It is important to remember that the interaction volume with the sample is broader than the beam diameter itself due to elastic scattering, thus, elemental information just adjacent to the beam can be often observed in the resulting spectrum. The beam can probe an area of the surface that is very small, less than one micron minimally; the beam can also be made very broad to collect information from a large sample area.

The following example is foreign material embedded in a polysilicon surface. Such defects in a surface can be problematic for MEMS that come into contact as this would act as an asperity, or in optical MEMS as scattering would occur. Defects like this can also be originators for stress concentrations. Thus, EDS was used to compare the materials in area 1 (defect) versus area 2 (background), in Fig. 5.27.

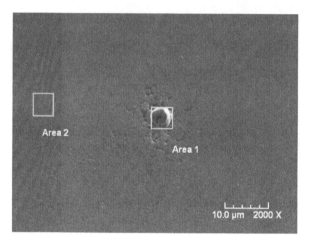

Fig. 5.27   SEM image of embedded particle in polysilicon

Area 1 and Area 2 have identical EDS spectra, as seen in Fig. 5.28(a, b). Thus, the source of the defect is silicon based, and is likely polysilicon as it was embedded in the polysilicon film. Polysilicon deposition tool related contamination is the root cause, and more frequency tool cleanings are corrective action for this defect.

5.5.9 Auger Analysis

Auger analysis is a technique which again uses an electron beam to excite the surface. The Auger electron is emitted which is much lower energy than the x-rays detected in EDS and is characteristic in energy for each element. Thus, the information gained is more surface sensitive as the Auger electron does not have a high enough energy to travel through and be emitted from the bulk of the sample. This technique can probe at a depth on the order of nanometers. The surface can be

Fig. 5.28 Embedded foreign material EDS spectrum matches background film. Area 1 from Fig. 5.26 is (**a**) while area 2 is (**b**)

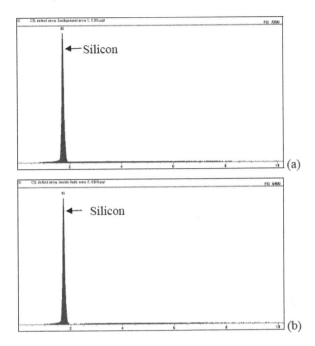

sputtered in the vacuum chamber and spectra can be collected upon completion of sputtering. Sputtering is performed sequentially to gain elemental data in a "depth profiling" manner. The beam can be reduced to less than one micron in diameter which allows for localized elemental surface information.

The Auger electron nomenclature is originated by the process of emission, and the illustration in Fig. 5.29 is of a silicon atom [18]:

1. The hole is the shell is in the K valence at electron energy level 1839.
2. It is filled by an L_1 electron while an L_{11}, L_{111} electron is emitted.
3. This is the Auger electron; nomenclature used is K L_1 L_{11}, L_{111}.

The example presented here is a silicon surface coated with gold and an adhesion layer of chromium [19]. Figure 5.30(a) shows an unannealed Auger depth profile spectrum, with a trace amount of copper at the very surface, and a small amount of oxygen that sits at the interface of the gold and silicon, with the chromium adhesion layer. After annealing for 225°C for 24 h, the depth profile changes to Fig. 5.30(b). Chromium and silicon have diffused to the surface and oxidized, while gold have diffused into the silicon. For reflective surfaces such as MEMS mirrors, the diffusion due to thermal effects seen here would be deleterious to reflectivity (Section 5.2, FMEA). The thermally-driven diffusion behavior of chromium though gold thin films and subsequent chromium oxidation has been observed in devices other than MEMS, such as quartz crystals [20]. Thus, to eliminate this deleterious diffusion

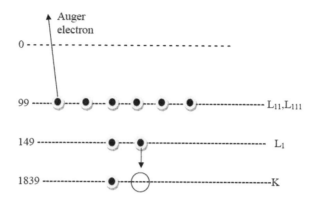

Fig. 5.29 Illustration of emission of Auger electron from an ionized silicon atom. Reference [18] reprinted with kind permission of Springer Science and Business Media, Copyright 1984

at temperature, an adhesion layer as well as a diffusion layer is recommended for MEMS mirrors that require subsequent thermal processing such as packaging [21]. Of course, careful choice of the thin films, process parameters to control deposition and their behavior over temperature must be considered to control curvature (Section 5.3.1).

5.5.10 X-Ray Photoelectron Spectroscopy (ESCA/XPS)

ESCA, also called XPS, is X-Ray Photoelectron Spectroscopy, a technique in which an x-ray beam probes the sample, detecting an electron emitted. The kinetic energy of the emitted electron is measured. The x-ray beam is of a known energy, thus, the binding energy of the emitted electron can be calculated (eq. 5.6). In this way, chemical information about the sample is generated.

$$\hbar\omega = E_{kin} + E_B^F(k) \tag{5.6}$$

In Eq. (5.6), $\hbar\omega$ is the x-ray beam energy (known), E_{kin} is the measured kinetic energy of the emitted electron, and $E_B^F(k)$ is the binding energy of the emitted electron. The "F" superscript means the binding energy is referenced to the Fermi level (for a solid) and the "k" is the level at which the electron is emitted. This technique can also utilize an ultra-violet (UV) energy source to probe the sample, and is called UPS.

In the following example, an anti-stiction coating was deposited using self-assembling monolayers based on tri-functional silanes. (See Chapter 3 for a complete description of the physics of failure of stiction in MEMS.) In this case, FOTS ($C_8H_4Cl_3F_{13}Si$) is first prepared for surface deposition through a silane hydroxylization step which gives off HCl. R in Eq. (5.7) is $C_8H_4F_{13}$, a partially fluorinated alkyl chain [22].

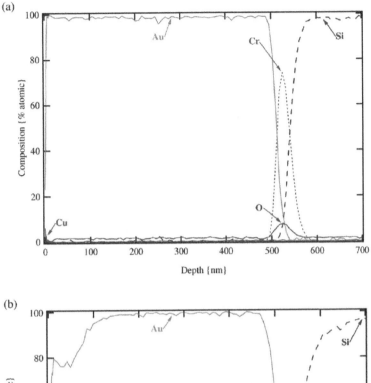

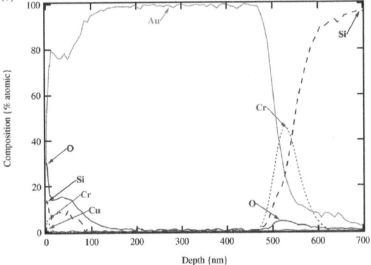

Fig. 5.30 (a) Auger depth profile prior to anneal, (b) Auger depth profile after anneal. Reference [19] reprinted with permission. Copyright 2007 Elsevier

$$R - SiCl_3 + 3H_2O \rightarrow R - Si - (OH)_3 + 3HCl \qquad (5.7)$$

The hydroxylized silane then bonds to a hydroxylized mono-crystalline silicon surface, and forms a low surface energy, hydrophobic film that is very thin. The thermal stability of SAM anti-stiction coatings are of concern due to thermal processing

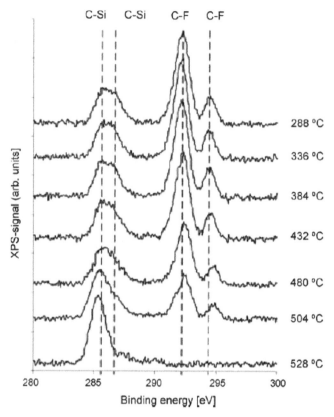

Fig. 5.31 XPS data of SAM coating on silicon at various temperatures. Reference [22] reprinted with permission. Copyright 2007 Elsevier

post SAM/anti-stiction coating, such as during packaging and PCB board mount operations. Here, XPS was used to determine the binding energy peaks of the SAM/silicon surface at various temperatures.

In Fig. 5.31, as the temperature increases, the C-F binding energy peaks reduce in signal starting at 400–450°C and are eventually gone at 528°C. The reduction of the C-F peak in the XPS data results in reduced static contact angle measurements, which indicates that the hydrophobic surface condition is lost at high temperature [22].

5.5.11 Time of Flight Secondary Ion Mass Spectroscopy (TOFSIMS)

Time of flight secondary ion mass spectroscopy (TOF-SIMS) is a very surface sensitive technique that has a depth resolution of 1–3 monolayers. A pulsed beam of primary ions (typically Ga+ or Cs+) impacts the sample surface, producing secondary ions through a sputtering process. The secondary ions are detected with mass spectrometry. Full periodic table elemental as well as molecular information can be obtained. Detection limits are 10^7–10^{10} atoms/cm^2 (sub-monolayer). This technique can be used to survey the surface, produce maps of the surface, and depth profile.

TOF-SIMS was used to survey a surface treated with an anti-stiction coating that was subsequently inhibited by deleterious organic deposition from outgassing organics in the packaging of a MEMS device. In the case presented, a silicone based contaminant adsorbed to the surface. Most silicones will outgas and the outgassing products can sit on the surface of an anti-stiction coating, inhibiting its effectiveness. Figure 5.32 is a TOFSIMS spectrum with silicone-based contamination inhibiting the anti-stiction properties of the fluorocarbon-based coating. The mass spectrometry technique identifies fragments of molecules based on their mass/charge ratio (m/z). The figure below identifies the fluorocarbon fragments using stars, and the silicon contamination using diamonds [23]. Elimination of this anti-stiction inhibition can be performed with better materials selection in packaging or by pre-conditioning organics to reduce outgassing [24].

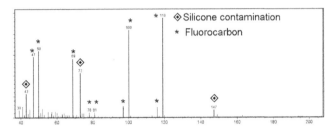

Fig. 5.32 Silicone contaminated MEMS surface originally treated with a fluorocarbon anti-stiction coating. Reference [23] reprinted with permission. Copyright 2007 Nano Science and Technology Institute

5.5.12 Fourier Transform Infrared Spectroscopy (FTIR)

Fourier Transform Infrared Spectroscopy is based on the absorption of infrared radiation. For covalently bonded materials, the energy of vibration or rotation is increased with infrared radiation absorption, as long as the dipole moment is changed during this vibration. This does not happen for the following dimers: H_2, O_2, N_2. Vibrations are termed stretching, bending (in plane or out of plane), rocking, etc. The spectrum for FTIR is plotted in absorption (or transmittance) versus wavenumber, as these vibrations occur at characteristic frequencies for specific bonds. This technique provides chemical information versus only elemental information like EDS or Auger analysis.

Polydimethylsiloxane (PDMS) airborne molecular contamination can absorb to an antistiction coating and inhibit it's effectiveness. The example for TOFSIMS

MEMS Surface	Work of Adhesion	Reference
Polysilicon Surface	• 100–300 mJ/m² • 20 mJ/m² with 58 Å surface roughness	• [25, 26] • [27]
Antistiction coating OTS (Octadecyltricholorsilane)	• 8 μJ/m²	• [28]
PDMS (polydimethylsiloxane)	• 20.7 mJ/m²	• [29]

Fig. 5.33 Work of adhesion for various surfaces, including PDMS

showed inhibition with a silicone based packaging material. PDMS is in many coatings and adhesives. Care should be taken to identify if PDMS has inhibited the MEMS antistiction coating, identify its source, and eliminate it. In Fig. 5.33, the work of adhesion is cited for several surfaces. Polysilicon is in the 20–300 mJ/m² range [25, 26]. Applying an anti-stiction coating such as OTS will reduce this surface parameter to single digit μJ/m². Yet adsorption of PDMS will greatly increase the work of adhesion, increasing the probability of stiction. The chemical structure of PDMS is shown in Fig. 5.34.

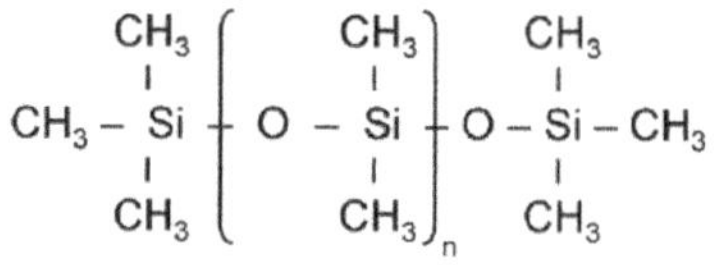

$$
\begin{array}{ccccc}
& CH_3 & \left(\begin{array}{c} CH_3 \\ | \end{array}\right. & & CH_3 \\
& | & & & | \\
CH_3 - & Si & - O - Si & - O - Si - & CH_3 \\
& | & & & | \\
& CH_3 & \left.\begin{array}{c} | \\ CH_3 \end{array}\right)_n & & CH_3
\end{array}
$$

Fig. 5.34 Chemical structure of PDMS. Reference [30] Copyright 2009 29th ICPIG, Cancun, Mexico

As it is present in packaging materials and is so deleterious to anti-stiction coatings, the infrared absorption spectrum of a thin film of PDMS is in Fig. 5.35. This was taken with infrared reflection absorption spectroscopy which consists of an FTIR with a special reflection unit that allows angle of incidence and polarization variation. In the PDMS spectrum, the CH_3 group has two C-H stretch vibration modes at 2965 cm^{-1} (symmetric) and at 2906 cm^{-1} (asymmetric). Wavenumber

Fig. 5.35 Infrared absorption spectrum of PDMS. Copyright 2009 29th ICPIG, Cancun, Mexico

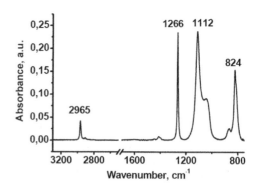

$1266~\text{cm}^{-1}$ is characteristic of Si–CH$_3$ (symmetric deformation). Rocking vibrations for Si–(CH$_3$)$_2$ and Si–(CH$_3$)$_3$ are observed at $824~\text{cm}^{-1}$ and $866~\text{cm}^{-1}$, respectively. Finally, the peaks at $1112~\text{cm}^{-1}$ and $1043~\text{cm}^{-1}$ are from asymmetric stretching vibrations of Si–O–Si. This example highlights the importance of having background spectra of known anti-stiction inhibitors.

5.6 Summary

Identification of root cause of failure is one of the most important aspects of MEMS reliability. Often, the MEMS engineer can be discovering new phenomenon specific to their device and application. Many failure mechanisms exist in the literature as well for both microelectronics and MEMS. Use of the Reliability FMEA to quantify field failure risks for evaluation is presented with an example of optical switch micromirror curvature. Modeling and accelerated testing are detailed for this mechanism. Historic data, burn-in, and failure analysis are also important to root cause identification. Use of the Think-Do-Check-Act method allows implementation of corrective action to be carefully performed.

References

1. Document SAE J 1739: Potential Failure Mode and Effects Analysis in Design (Design FMEA) and Potential Failure Mode and Effects Analysis in Manufacturing and Assembly Processes (Process FMEA) Reference Manual, SAE, 400 Commonwealth Drive, Warrendale, PA 15096–0001.
2. Bhattacharya, S., Hartzell, A. (2007) J. Micro/Nanolith, MEMS MOEMS, Jul–Sep 6(3), 033010-1–033010-12.
3. Stoney, G.G. (1909) The tension of metallic films deposited by electrolysis. Proc. R. Soc. London, Ser. A. 82(553), 172–175.
4. LaVern, A.S. (2002) PhD Thesis, Air Force Institute of Technology, April 2002 "Characterization of Residual Stress in Microelectromechanical Systems (MEMS) Devices using Raman Spectroscopy".
5. Ken, G., et al. (2004) Creep of thin film Au on bimaterial Au/Si microcantilevers. Acta Materialia 52, 2133–2146.
6. Arthur Lin, Y. (1999) Parametric Wafer Map Visualization. IEEE Comput. Graphics Appl. 19(4), 14–17, (Jul/Aug).
7. Arman G., et al. (2000) Mechanical Reliability of Surface Micromachined Self-Assembling Two-Axis MEMS Tilting Mirrors. Prov. SPIE. 4180, MEMS Reliability for Critical Applications.
8. Wyko NT9100 Optical Profiling System, 2007 Veeco Instruments Inc. DS544, Rev A0.
9. Koev, S.T., Ghodssi, R. (2008) Advanced interferometric profile measurements through refractive media. Rev. Sci. Instrum. 79, 093702.
10. Goldstein J, Newbury DE, Joy DC, Lyman CE (2003) *Scanning Electron Microscopy and X-ray Microanalysis*. New York: Springer.
11. Kahn, H., Ballarini, R., Heuer, A.H. (2001) On the Fracture Toughness of Polysilicon MEMS Structures. Mat. Res. Soc. Symp. Proc. 657 (© 2001 Materials Research Society). 13–18.

12. Miller, D.C., et al. (2008) Connections between morphological and mechanical evolution during galvanic corrosion of micromachined polysilicon and monocyrstalline silicon. J. Appl. Phys. 103, 123518.
13. Guy F. Dirras, George Coles, Anthony J Wagner, Stephen Carlo, Caroline Newman, Kevin J. Hemker, William N. Sharpe, "On the Role of the Underlying Microstructure on the Mechanical Properties of Microelectromechanical Systems (MEMS) Materials" Materials Science of Microelectromechanical Systems (MEMS) Devices III, MRS Proceedings Volume 657.
14. Nunan, K., Ready, G., Sledziewski, J. (2001) LPCVD and PECVD Operations Designed for *i*MEMS Sensor Devices. Vacuum Coating Technol. 2(1), 26–37.
15. Gnauck, P., Hoffrogge, P. (2003) A new SEM/FIB Crossbeam Inspection Tool for high Resolution Mateirals and Device Characterization. Proc of SPIE. 4980, Reliability, Testing, and Characterization of MEMS/MOEMS II.
16. Walraven, J., et al. (2000) Failure analysis of tungsten coated polysilicon micromachined mircroengines. Proc. of SPIE. 4180, MEMS Reliability for Critical Applications.
17. Bharat, B., Huiwen, L. (2004) Micro/nanoscale tribological and mechanical characterization for MEMS/NEMS. Proc. of SPIE. 5392, Testing, Reliability and Application of Micro- and Nano-Material Systems II.
18. Loretto, M.H. (1984) *Electron Beam Analysis of Materials*. New York: Springer Science and Business Media.
19. Miller, D., et al. (2007) Thermo-mechanical evolution of multilayer thin films: Part II. Microstructure evolution in Au/Cr/Si microcantilevers. Thin Solid Films 515, 3224–3240.
20. Thornell, G., et al. (1999) Residual stress in sputtered gold films on quartz measured by the cantilevel beam deflection technique. IEEE Trans Ultrasonics, Ferroelectrics, Frequency Control, 46(4), July.
21. Alie, S., Hartzell, A., Karpman, M., Martin, J.R., Nunan, K. (2003) Optical mirror coatings for high-temperature diffusion barriers and mirror shaping United States Patent 6508561, Analog Devices.
22. Knieling, T., Lang, W., Benecke, W. (2007) Gas phase hydrophobisation of MEMS silicon structures with self-assembling monolayers for avoiding in-use sticking. Sensors Actuators B 126, 13–17.
23. Mowat, I., et al. (2007) Analytical methods for nanotechnology. NSTI Nanotech 2007 Proceedings, Santa Clara, May 20–24.
24. Tepolt, G.B. (2010) Hermetic vacuum sealing of MEMS devices containing organic components. SPIE 2010 Reliability, Packaging, Testing, and Characterization of MEMS/MOEMS and Nanodevices IX, Conference 7592.
25. Mastrangelo, C.H. (1999) Supression of Stiction in MEMS. MRS.
26. Mastrangelo, C.H., Hsu, C.H. (1992) A simple experimental technique for the measurement of work of adhesion of microstructures. Solid-State Sensor and Actuator Workshop, 1992, 5th Technical Digest, IEEE; 22–25 June.
27. Maboudian, R., Carraro, C. (2004) Surface chemistry and tribology of MEMS. Ann. Rev. Phys. Chem. 55, 35–54.
28. Wibbeler, J. et al. (1988) Parasitic charging of dielectric surfaces in capacitive microelectromechanical systems (MEMS). Sensor Actuators A 71, 74–80.
29. Reiter, G. et al. (1999) Destabilizing effect of long-range forces in thin liquid films on wettable surfaces. Europhys. Lett. 46(4), 512–518.
30. Danilov, V. et al. (2009) Plasma treatment of polydimethylsiloxane thin films studied by infrared reflection absorption spectroscopy. 29th ICPIG, July 12–17, Cancun, Mexico.

Chapter 6
Testing and Standards for Qualification

6.1 Introduction

It can be said that the semiconductor is based on three elements: the transistor, the capacitor and the wire. There is not a standard set of structures in the MEMS world; this is a very versatile field with little boundaries and new products developing daily (bioMEMS from polymers, inertial MEMS, powerMEMS, optical MEMS, RF MEMS, etc.). Thus, testing each MEMS product type and design can require unique instruments that are often custom designed. Development of the product itself and test platforms that quantify the test distribution of MEMS parts is critical to produce the product. If the part test distributions all fall within the production specification then 100% yield is achieved, the ultimate goal for any manufacturing line (Chapter 7).

One can say the same for qualification of MEMS products. Qualification is a set of tests, typically run to an industry standard based on the customer's application that assures the MEMS design and process flow are robust enough for the product to transition into production mode. Qualification standards for MEMS are an interesting but complex topic. Because of the myriad of structures and applications, operating environments and storage can be very different as are physics of failure (Chapter 4). There has been plenty of discussion on MEMS standards, but no defined set of standards to date exists.

6.2 Testing MEMS

A thorough understanding of acceptable and destructive operating conditions in both the field and in stress testing is found through the study of distribution data. Figure 6.1 is an illustration of how distributions are used for setting production specifications, operating margins, and destruction limits as a function of stress, such as mechanical testing [1]. The full range of operation is here: from destruction limits (where not to operate), to upper and lower operating limits (where the part will operate without destruction) and product specification (the area in which the customer will use the product). The area outside of the product specification where operation

A.L. Hartzell et al., *MEMS Reliability*, MEMS Reference Shelf,
DOI 10.1007/978-1-4419-6018-4_6, © Springer Science+Business Media, LLC 2011

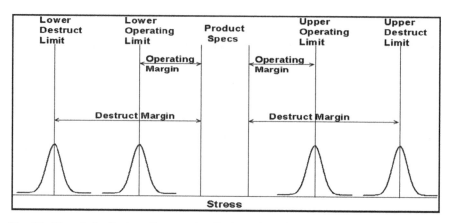

Fig. 6.1 Use of distributions for operating limits and product specifications. Reference [1] reprinted with permission. Copyright 2008 SPIE

is still not destructive is termed the operating margin. MEMS manufacturers limit the production specification to a safe area within the operating limits, far away from the destruct limits. Setting these ranges can be performed with a variety of data collection methods.

Development of test structures and short loop process runs are key for early data generation (Chapter 7). Empirical data gathered by multiple test structures is cost effective, but only if the test structures are representative of the product. Short loop process runs are defined as shorter cycle time process snapshots that allow rapid cycles of learning. Testing of production parts that run through the entire process flow is important to perform in parallel with the data collection from test structures and short loops. Comparison of test data from all three methods can assure the test structures and short loop experiments are proper predictors of the actual product performance. This combination, done properly, can reduce the time to market of a new MEMS product by improving yield and therefore reliability.

The MEMS engineer can analyze product distributions through proper test methodologies. If the product has an unacceptable failure rate, root cause and failure analysis techniques are used (Chapter 5) to identify the physics of failure and implement corrective action. This type of iterative exercise will result in a product distribution that meets production specs, with high confidence in field operation behavior through data from accelerated testing (Chapter 2) and qualification.

6.2.1 Classes of MEMS Devices

Zunino and Skelton [2] created the concept of MEMS device classes that are based on characteristics that are primarily mechanical in nature. Table 6.1 is their conceptual classification. Note that the term 'no moving parts' means no parts move continuously in operation. The accelerometer does have moving parts yet

Table 6.1 Classification of MEMS by characteristics and examples from the US Department of Defense. Reference [2] reprinted with permission. Copyright 2006 SPIE

MEMS classifications		
Class of MEMS	Characteristics	Examples
Class I	No moving parts	Accelerometers Pressure Sensors
Class II	Moving parts; no impacting surfaces	Gyros Resonators Filters
Class III	Moving parts; impacting surfaces	Relays Valves Pumps

Class IV	Moving parts;
	  impacting and rubbing
	  surfaces
		Optical switches
		Scanners
		Discriminators

Class V	Moving parts; interfaces
	  with explosives,
	  propellants & energetic
		S & A
		IMUs
		Fuzing

is stationary during operation until it receives an external mechanical shock; upon receipt of this shock the accelerometer beams move and an output proportional to the movement is detected in the system housing the accelerometer (Chapter 2). This would be termed a Class I device in the Table 6.1. In addition, van Spengen [3] published a list of generic MEMS elements in by defining structures (Table 6.2). These examples of classification can be useful in defining a test and qualification program for MEMS. Some MEMS applications such as Microfluidics are not included in these charts yet for qualification purposes, they would still have to pass the qualification standards that are required by the customer and the application.

Testing MEMS includes two major areas: characterization and production screening which are both required to characterize designs and structures, as well as develop test methods for production. Characterization testing includes quantifying the frequency modes and Q, for example, of the MEMS structural elements by a laser Doppler vibrometer.

6.3 Test Equipment for MEMS

MEMS are tested in the same methodology as testing in the semiconductor field. Primarily, testing is performed at wafer level and at packaged MEMS level. Testing is also performed at intervals for reliability and qualification testing. Traditional semiconductor test equipment is used often for electrical probing of MEMS wafers and for testing packaged devices. This section will highlight examples of unique test

Table 6.2 Generic MEMS elements. Reference [3] reprinted with permission. Copyright 2003 Elsevier	Generic MEMS elements

	• Structural beams
	  ○ Rigid
	  ○ Flexible
	  ○ One side clamped
	  ○ Two sides clamped
	• Structural thin membranes
	  ○ Rigid
	  ○ Flexible
	  ○ With holes
	• Flat layers (usually adhered to substrate)
	  ○ Conductive
	  ○ Insulating
	• Hinges
	  ○ Substrate hinge
	  ○ Scissors hinge
	• Cavities
	  ○ Sealed
	  ○ Open
	• Gears
	  ○ Teeth
	  ○ Hubs
	• Tunneling tips
	• Reflective layers

equipment that has been adopted or developed for the MEMS industry. Some commercially available test platforms are available (covered in Sections 6.3.3–6.3.5). Examples of custom equipment are shown in Sections 6.3.1, 6.3.2, and 6.3.6.

6.3.1 Shaker Table for Vibration Testing

Mechanical structures, even Class I devices, will vibrate with externally applied vibration and shock. Thus, the shaker table (manufacturers include Unholtz Dickie, M/Rad Corporation, ETS Solutions) is a common test platform used to characterize and test the response of MEMS structures to outside stimulus; they are used during production test, acceleration testing, and qualification testing. Vibration modes from an electrodynamic shaker table are controlled with a power amplifier and DC power source. Vibration modes simulate the exposure the part will have in operation, post-packaging production, and for qualification testing (see Section 6.4.1.3). The shaker table system is a platform to provide vibration and can be integrated with externally supplied test fixtures, electronics, accelerometers, and unique equipment required for the specific MEMS structure and application.

The set up in Fig. 6.2 consists of a fixture that rigidly mounts packaged MEMS devices to the shaker table as well as a balance mass that prevents unwanted out of plane movement. The fixture can be mounted in various orientations. Figure 6.3 has

Fig. 6.2 Shaker table with arrow indicating axis of movement. Reference [4] reprinted with permission. Copyright 2000 IEEE

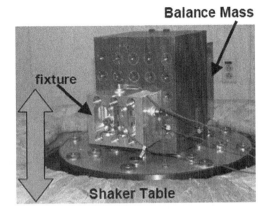

Fig. 6.3 Power spectral density operating environment (termed requirement) and accelerated conditions (termed test) for shaker table testing. Reference [4] reprinted with permission. Copyright 2000 IEEE

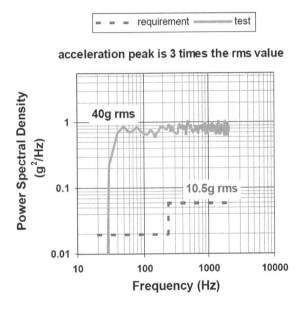

both the operational requirement and accelerated test conditions for the packaged MEMS device in a power spectral density (g²/Hz) plot that is a function of vibration frequencies. Note the testing was done from 20 Hz to 2000 Hz (Section 6.4.1.3). To calculate gee-level RMS values, equation (6.1) was used in reference [4].

$$RMS(g) = \sqrt{(PSD * Bandwidth)} \qquad (6.1)$$

In this case the packaged MEMS structure is the Sandia microengine (covered in Chapter 2 in failure distributions). This complex device includes a shuttle that has

Fig. 6.4 Elements of the
Sandia Microengine, with
higher magnification images
showing the shuttle and gear
and pin joint. Reference [4]
reprinted with permission.
Copyright 2000 IEEE

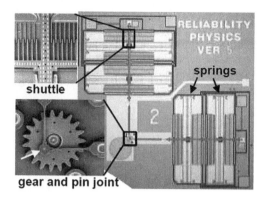

moveable beams and a gear and pin joint. The gear has teeth and a hub (Table 6.2)
while the pin joint joins the linkage arms to the gear. The microengine is classified
as a Class IV device (Table 6.1; Fig. 6.4).

The shuttle is a comb drive structure in which electrostatic stimulus results in
movement of the two sides of the shuttle relative to one another. Debris was present
beneath the shuttle prior to the vibration test. The debris moved during the testing,
but did not create a failure (Fig. 6.5). If the debris became stuck in the comb drive,
the shuttle would mechanically obstruct and no longer move. Debris that is electrically conductive could create a short between the combs themselves or the comb
and ground structure beneath. Additional detail on application of vibration testing

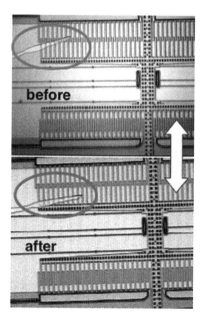

Fig. 6.5 Example of foreign
material movement in the
comb drive area of the
microengine, as a result of the
shaker testing, yet no failure
was observed in this case.
Reference [4] reprinted with
permission. Copyright 2000
IEEE

is covered later in this chapter and in Chapter 4. Contamination related failure is covered in Chapter 3.

6.3.2 Optical Testing for Deformable Mirrors

Optical MEMS devices require special test platforms (Chapter 2 shows the Texas Instruments DMD test station). The MEMS deformable mirror (DM) for adaptive optics (AO) is required for high performance diffraction limited imaging in applications such as astronomy and ophthalmology. Adaptive optics is used to compensate (in real time) for wavefront aberrations caused by the propagation of light through turbid media, such as the atmosphere or eye tissue. The primary components of the AO systems include a wavefront sensor and a DM. The DM corrects the phase of the aberrated wavefront in a closed loop configuration using integrated control electronics to drive the mirror actuators during operation. AO kits are now commercially available from Thorlabs [5], yet test of the MEMS DM itself is required to assure quality performance prior to integrating it into the system. This testing is performed on a laboratory test bed.

Figure 6.6 is a schematic of such a MEMS DM test bed [6]. A laser diode enters a single mode fiber, the output of which is collimated by the lens f1 at a specific beam diameter. This beam is Gaussian and the iris blocks all but the center portion

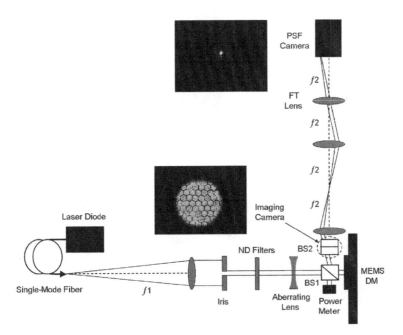

Fig. 6.6 Optical test bed for far field diffraction test of deformable mirror (DM) in adaptive optics applications. Reference [6] reprinted with permission. Copyright 2006 IEEE

of the Gaussian beam. A beam splitter (BS1) takes 50% of the light and directs it to a power meter, which measures the light intensity (for Strehl ratios), and the other 50% is reflected off the MEMS DM and directed to an imaging camera in the test bed. The DM is aligned using reflected light from the 2nd beam splitter (BS2) and the imaging camera; the reflected light moves through another series of lenses and to a 2nd imaging camera termed the PSF camera which measures the deformable mirror far field diffraction, also known as the point spread function (PSF). A multi-mirror with flat segments (flatness from manufacturing processing) will result in the best PSF. A packaged tip-tilt piston DM is shown in Fig. 6.7. Mirror segment flatness is measured with interferometry (Section 5.5). For example, the control electronics can individually displace each segment of the mirror to create an overall curved structure of the MEMS multi-mirror for defocus (Fig. 6.8). MEMS DM optical mirror designs are available in Classes II, III, and Class IV.

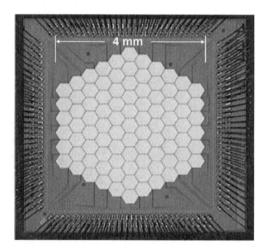

Fig. 6.7 Segmented MEMS DM packaged multi-mirror device. Reference [6] reprinted with permission. Copyright 2006 IEEE

Fig. 6.8 MEMS DM mirror with no applied voltage on left, voltage is applied to DM on right to defocus system. Reference [7] reprinted with permission. Copyright 2008 SPIE

6.3.3 Dynamic Interferometry

MEMS structures that move during operation require test platforms that measure their dynamic behavior, as some failure mechanisms cannot be observed when the part is in stationary mode. Veeco Wyko has commercially available systems that consist of a strobed light source that can measure MEMS in motion. Movement of structures results in changes to the interferogram. The addition of strobed illumination operating at the frequency of MEMS drive movement can result in dynamic measurement capability. Veeco's system design includes extensive software and proper LED and CCD framing and exposure control, which allows generation of a series of images in time. Together, these images can be presented in a movie-like fashion to capture dynamic behavior. In-plane and out-of-plane motion, deflection, rotation, distortion, radius of curvature, bow, and range of motion are amongst the parameters that can be measured while the MEMS device is in motion. Figure 6.9 is a phase versus displacement plot and 3D map of an in-plane MEMS resonator (Class II device). Veeco calls this their In-Motion^{TM} feature, which can measure at frequencies up to 2.4 MHz [8]. Figure 6.10 is a picture of the system.

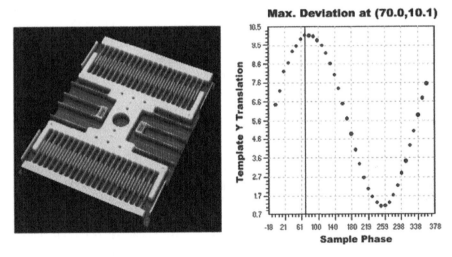

Fig. 6.9 Phase versus displacement plot – data analysis of in-plane resonator with 3D map [8]. Courtesy of Sandia National Laboratories, SUMMiT(TM) Technologies, www.mems.sandia.gov

6.3.4 MEMS Optical Switch Production Test System

Polytec and Cascade teamed together to combine a probe station with a Doppler laser vibrometer; these commercially available systems were designed to provide non-destructive device characterization for MEMS optical switch devices. This test platform is the Series 12,000 Switch Resonance system. It consists of the Cascade

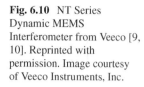

Fig. 6.10 NT Series
Dynamic MEMS
Interferometer from Veeco [9,
10]. Reprinted with
permission. Image courtesy
of Veeco Instruments, Inc.

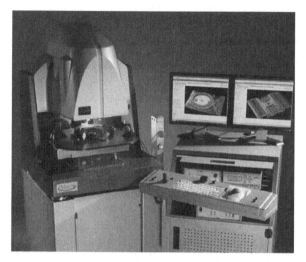

Summit 12,000 Series Prober with high and low temperature capability, a Polytec
Micro-Scanning Laser Doppler Vibrometer (LDV), a vibration isolated environment
and Cascade Microtech software [11]. The system is shown in Fig. 6.11.

The Laser Doppler vibrometer itself is a very powerful technique that provides
dynamic information of MEMS devices. In LDV, the MEMS device is driven with a
periodic AC drive, and its out-of-plane velocity is measured as a function of time by
detecting the frequency shift of a reflected laser beam. A Fourier transform can
provide acceleration, velocity, or displacement amplitude vs. frequency, all as a
function of time. The resonance frequencies of different modes and determination
of which modes are excited can be detected and recorded. The laser vibrometer also
provides animation of recorded data: an adjustable mesh can cover a single device
or entire array, and the laser beam of the laser vibrometer scans set-points within
the mesh and records data from different devices within a few seconds. The interferometric technique allows monitoring of MEMS device motion with sub-nanometer
resolution (Fig. 6.12).

The addition of the Cascade prober allows drive voltage through microprobe
application, allowing individual MEMS structures to actuate while obtaining non-
contact LDV measurement of optical switch characteristics such as deflection
amplitude, settling time, resonant frequency, and cross talk between MEMS optical switches. Figure 6.13 shows a probed MEMS optical switch test vehicle [12].
The Lucent optical switch is covered in Chapter 4 Lucent's optical switch is a Class
II device.

6.3.5 Laser Doppler Vibrometer/Strobe Video System

Other manufacturers offer hybrid systems. For instance the Polytec MSA-500 com-
bines scanning laser Doppler vibrometry, White light interferometry, and strobed

Fig. 6.11 MEMS Optical
Switch Production Test
System with Cascade
Microtech Probe Station [11].
Reprinted with permission.
Courtesy Polytec GmbH,
Cascade

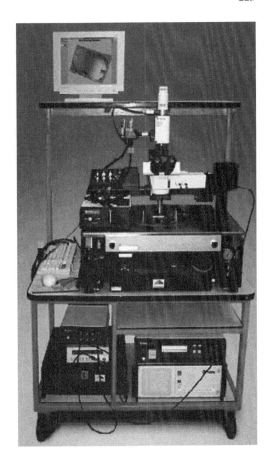

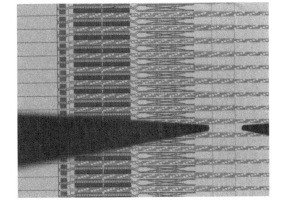

Fig. 6.12 Probe needles on
the contact pads of an array of
160 μm wide silicon
micromirrors. Courtesy
W. Noell at the EPFL and
S. Weber at the University of
Geneva [12]

Fig. 6.13 Polytec MSA-500, on a Cascade Microprobe Test Station. Reprinted with permission. Courtesy Polytec GmbH

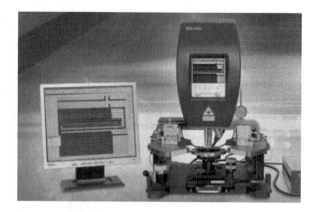

in-plane vision recognition. This system provides in-plane motion analysis by adding a stroboscopic video system to the LDV that synchronizes LED flash time and camera on-time to the excitation of the MEMS structure [13].

A comb-drive MEMS structure is measured by a Polytec with strobed in-plane vision recognition. The resonant frequency is first measured with out-of-plane motion by the Polytec Laser Doppler Vibrometer, and then the in-plane stobed mode is switched on to characterize the sensitivity (displacement versus voltage) of the structure as it operates in plane. This technique can be used for production testing, interval testing during acceleration and qualification testing, and for characterization. The results are shown in Fig. 6.14.

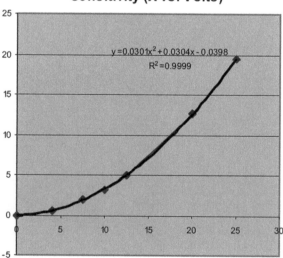

Sensitivity (X vs. Volts)

$y = 0.0301x^2 + 0.0304x - 0.0398$

$R^2 = 0.9999$

Fig. 6.14 In-plane comb-drive sensitivity measurement [13]. Reprinted with permission. Copyright 2004 SPIE

6.3.6 *SHiMMer (Sandia High Volume Measurement of Micromachine Reliability)*

Sandia has developed *SHiMMer* (Sandia High volume Measurement of Micromachine Reliability) to control and measure up to 256 microengines at once while in a controlled environment. This custom test platform is used to generate reliability data on multiple parts at once, and to create lifetime distribution data. Data from this device as tested on the SHiMMer is shown in the lifetime statistics sections in Chapter 2. As can be seen in Fig. 6.15, this tool consists of electronics (waveform generation, amplification, and synthesis), an X/Y positioner with a video microscope, camera, video diagnostics and recording, and reliability testing software. Parts under test sit on the X/Y positioned table, and are packaged in test sockets to receive electrical stimulus. As the microengine's linkage arms will move through comb drive actuation to drive a revolution of the gear, the optical imaging of the system can detect a failure such as a stuck gear or anomalous oscillations of the gear.

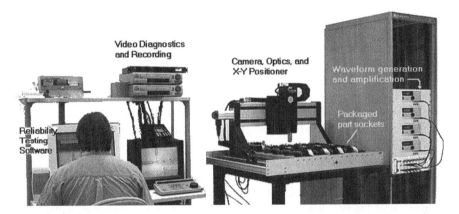

Fig. 6.15 SHiMMer System at Sandia Laboratories. Reference [14] reprinted with permission. Copyright 1997 SPIE

6.4 Quality Standards and Qualifications

Qualification testing of MEMS starts with a discussion between the customer and the MEMS supplier. The application of the final system will dictate the qualification standard, yet these standards were developed for semiconductor devices and other parts – not MEMS. This up-front communication is therefore critical for the MEMS supplier, as choice of a test in a standard that is destructive and doesn't simulate the use or storage environment for the MEMS product can halt introduction of the part into the marketplace. Replacement with a suitable test instead of a destructive one can be made in the qualification discussion phases. The tests in Table 6.3

Table 6.3 Typical DMD test storage environments. Reference [1] reprinted with permission. Copyright 2008 SPIE

Qualification test	Test description	Hours/cycles (minimum)
Storage Life Cold/Hot	−55/100°C, no power	1000 h
Temperature Cycle	−55/125°C, air-to-air, fine/gross leak	1000 cycles
Thermal Shock	−55/125°C, liquid-to-liquid	200 cycles
Sequence 1	1500 g Mechanical Shock, Y only Vibration 20 g, 20–2000 Hz Constant Acceleration, 10Kg, Y1 only	
Sequence 2	Thermal shock, −55/125°C Temperature Cycles, −55/125°C Moisture Resistance	15 cycles 100 cycles 10 days

are temperature, humidity and mechanical tests for non-operating environment. In operation, the DMD is a Class IV device as classified in Fig. 6.2.

Qualification testing and equipment for both operating and non-operating conditions will next be covered qualitatively, and popular qualification standards' requirements will be covered in Sections 6.4.1–6.4.4. This section is not written to be all-inclusive, but instead, is to educate the reader and cover various test methods that can later be reviewed in depth by the MEMS reliability engineer.

High Temperature Storage is a simulation of elevated temperature non-operating conditions. This number is chosen by both the qualification standard and application. For instance, automotive high temperature storage conditions are a function of the position in the automobile: passenger compartment, under hood, or under hood next to the engine. Elevated temperatures can be experienced in many ways including inventory storage in warehouse environments that typically don't have HVAC for air conditioning, transportation in trucks on a hot day, even in airplane cargo holds when sitting on a heated runway. A timeframe of 1000 or 2000 hours are typically required for high temperature storage to assure that the MEMS part can survive a lengthy period at high temperature. Testing is performed at intervals that are defined in the standard or by the MEMS supplier. Physics of failure accelerated by elevated temperatures are covered in Chapter 4. (Fig. 6.16 is a single chamber system that can sustain high temperatures for storage testing.) This testing can also be performed under operation, with the proper electrical and/or mechanical stimulus in the chamber and either in-situ monitoring or with testing at intervals; this is termed operational qualification high temperature testing. See Chapters 2 and 4 for acceleration models that have both temperature and voltage terms for physics of failure.

Fig. 6.16 Temperature
CT-series chamber [15].
Reprinted with permission.
Courtesy, Cincinnati
Sub-Zero

Low Temperature Storage is a simulation of low temperature environments in non-operating conditions. Again this is application and qualification standard specific. Examples of low temperature excursions can be storage in an unheated warehouse in the cold winter, unheated airplane cargo holds where parts are shipped for extended periods at high elevation, and unheated transportation. Timeframes are also in the 1000 to 2000 hour range. The chamber in Fig. 6.16 can also hold stable cold temperatures for low temperature storage. Bench-top systems are also available and an example is in Fig. 6.17. Cooling can be boosted with liquid nitrogen in some chambers, while others have a dual cascade system of compressors that provides cooling. Low temperature operational testing is performed by applying electrical and/or mechanical stimulus to the device under test in the chamber. Testing of the qualification device is performed with in-situ or interval testing.

Fig. 6.17 Benchtop oven that
provides elevated and cold,
Microclimate Series [16].
Reprinted with permission.
Courtesy, Cincinnati
Sub-Zero

Temperature Cycling is self-defined, and the variables are the high temperature, the dwell at high temperature, the low temperature and dwell there, and the ramp rates between the temperature extremes. Some temperature cycling is done with very fast ramp rates using air-to-air shock chambers [17]. This equipment has two chambers, one is always hot, and one is always cold. An elevator type mechanism moves the parts between the two chambers for a very fast ramp rate. In a single chamber thermal cycle tool, the chamber itself changes temperature and therefore this ramp rate is much slower. Some failure mechanisms are tested with a fast ramp rate, and some with a slow ramp rate. Chapters 3 and 4 cover physics of failure accelerated by this testing, such as coefficient of thermal expansion mismatch and fracture mechanisms. Cycling is required for various conditions, ramp rates and number of cycles as a function of the qualification standard. The various temperature cycling conditions simulate environmental conditions of hot to cold, and cold to hot. A MEMS part under the hood of a car positioned near to the engine will be cold on a winter day in a parking lot, and will quickly heat up upon ignition at a fast ramp rate (Fig. 6.18).

Fig. 6.18 Air-to-air shock dual chamber system, VT series [18]. Reprinted with permission. Courtesy, Cincinnati Sub-Zero

Thermal shock is performed by physically moving the MEMS parts between two separate liquid baths: one held at elevated temperature and one held at very low temperature. Due to the thermal properties of the liquid, the ramp rate between temperature extremes is the highest, and therefore delivers the highest stress to the part. This again simulates thermal extremes and can bring out failure mechanisms such as glass frit package seal cracking, or ceramic fracture. This test is not always performed and is again a function of the discussion between the customer and MEMS manufacturer in the early stages of the product development (Fig. 6.19).

Temperature/Humidity testing is performed in either constant temperature, constant relative humidity conditions, or as a Moisture Resistance test with thermal cycling and humidity cycling. The former is designed to never hit the dew point for moisture condensation, while the latter can experience dew point conditions. The

Fig. 6.19 Liquid-liquid
thermal shock system with
two baths, TSB series [19].
Reprinted with permission.
Courtesy, Cincinnati
Sub-Zero

chamber in Fig. 6.16 can be fit with a humidification system and humidity controller
to become a temperature/humidity chamber to test humidity driven failure mechanisms such as corrosion. Galvanic corrosion mechanisms do not need an applied
voltage, yet with an applied voltage, mechanisms such as anodic oxidation can be
tested (Chapter 4).

Mechanical Shock testing is covered in detail in Chapter 4. Various test equipment can be used to deliver a mechanical shock. Shown here in Fig. 6.20 is a
simple guillotine tester. Mechanical shocks are experienced in both operating and
non-operating environments, and the part under test in the mechanical shock equipment can be powered or unpowered. Shock can occur in post processing, such as
a MEMS device that is robotically picked up and positioned into a printed circuit
board. Shipment and transportation deliver shocks to MEMS devices and systems.
The use application could be in an environment that experiences mechanical shock.
Thus, for mechanical structures, mechanical testing is extremely important.

Vibration testing was covered in an example in Section 6.3.1 of this chapter
with shaker table testing. Vibration occurs in both operational and non-operational
settings at various frequency and amplitude ranges. Again, shipping and storage
conditions will have vibration. Operating environments will have vibration; the

Fig. 6.20 Guillotine tester
for mechanical shock testing.
Reference [20] reprinted with
permission. Copyright 2006
SPIE

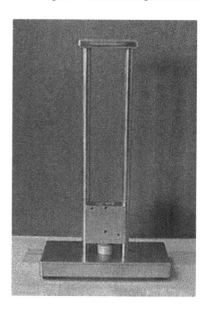

MEMS housing must be analyzed for damping and tested to the proper specifications. Severe vibration such as NASA mission launch will be discussed in
Section 6.5.

Constant Acceleration is a test developed to expose the part to extreme gee
level exposure in specific orientations. These are high levels but can highlight, as an
example, if the die attach material and process will hold the MEMS die properly in
the package. This test has some controversy associated with it, as it does not have
an acceleration factor nor does it test real field conditions for typical applications
and was designed to test structural element integrity of the device. If parts pass, they
show a robust design, material choices and manufacturing process. Constant acceleration conditions will be covered in Section 6.4. An example as constant acceleration
testing as a use condition is in Section 6.5.2.

Hermeticity is not a stress test, but is a method to test if a hermetic package has
kept its hermetic integrity. It is used in qualification testing post stress test exposure that could create a hermeticity failure. Fine and Gross leak methods are defined
in detail in Mil-Std-883. The industry standard minimum leak rate for semiconductor devices is 5×10^{-8} atm-cc/s, while the maximum water concentration in the
package is 5000 ppm. Previously mentioned was the dependence of corrosion mechanisms on relative humidity. The following relationship is used to develop the set
of curves in Fig. 6.21, where $p_{H_2O}^o$ is the saturation vapor pressure and p_s is 1 atm.
Equation (6.3) is calculated (temperature dependence), inserted into equation (6.2),
and then put into equation (6.1a) (RH dependence). Here, p is the atmospheric pressure, which in the case is the cavity pressure. Our example is for a cavity pressure
of 1 atmosphere [21].

RH vs temp for various water concentrations at cavity pressure=1 atm

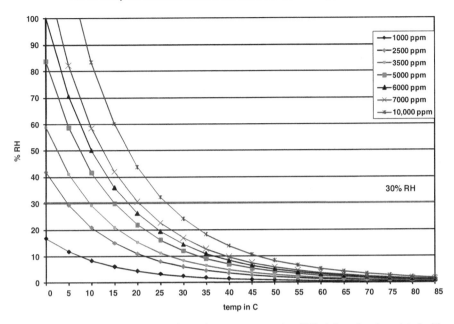

Fig. 6.21 Temperature versus humidity and water concentration [22]. Adapted and reprinted with permission. Copyright 2010 SPIE

$$[H_2O] = 10^4 RH \frac{p^o_{H_2O}}{p} (\text{ppm}) \qquad (6.1a)$$

$$p^o_{H_2O} = p_s \exp(13.3185a - 1.9760a^2 - 0.6445a^3 - 0.1299a^4) \qquad (6.2)$$

$$a = 1 - \frac{373.15}{T(K)} \qquad (6.3)$$

A full set of curves, varying RH and temperature, are plotted in Fig. 6.21. The case can be made for a lower maximum hermetic cavity water concentration for MEMS, as relative humidity-driven physics of failure can start to occur when just a few monolayers of water build up on the surface to act as an electrolyte. If one wants a generous safety factor and designs the maximum internal cavity RH to 30%, water concentrations at low temperatures must be greatly below 5000 ppm. This curve is generated for a low temperature of 0°C, yet these curves can be generated to much lower operating temperatures.

There are many methods to test for hermeticity. They are covered in Mil-Std-883H Method 1014.13 (Seal testing).

6.4.1 Mil-Std-883 (Revision H is current)

Mil-Std-883 (Revision H is current) is an industry standard used widely for semiconductor qualification testing. It is titled "Test Method Standard: Microcircuits".
The standard was originally written to provide uniform methods for qualification testing of microdevices, for military and aerospace applications. It contains
numerous test methods under each of the following categories:

- Environmental tests
- Mechanical tests
- Electrical tests (digital)
- Electrical tests (linear)
- Test procedures

Temperature ranges for high temperature storage, variable frequency vibration
levels, mechanical shock, and constant acceleration will covered from Mil-Std-
883H. Every MEMS reliability engineer should have a copy of this 729 page
standard. Some test conditions included in Table 6.3 are Mil-Std-883 conditions.

6.4.1.1 Temperature ranges from Stabilization Bake, Method 1005.9

The stabilization bake is a non-operating test to determine the effect of elevated
temperature storage on parts. Table 6.4 has the conditions and temperatures of
stabilization bake.

Table 6.4 Test conditions
and temperatures for
stabilization bake from
Mil-Std-883H

Test condition	Temperature (minimum)
A	75°C
B	125°C
C	See Table 6.5
D	200°C
E	250°C
F	300°C
G	350°C
H	400°C

6.4.1.2 Temperature cycling, Method 1010.8

Temperature cycling is a non-operating test. Temperature ranges are in Table 6.6.
The cycling can start with either hot or cold testing. The cycling profile for
Condition C is shown in Fig. 6.22. Note the ramp rates are not defined, but within
the method there is information about transfer rates for both dual chamber and single
chamber systems.

Table 6.5 Test condition C, temperature and duration, Method 1005.9, Mil-Std-883H

Minimum temperature (°C)	Minimum time (h)[a]
100[b]	1000
125[b]	168
150	24
155	20
160	16
165	12
170	8
175	6
200	6

[a] Equivalent test condition C duration
[b] These time-temperature combinations may be used for hybrid microcircuits only.

Table 6.6 Temperature ranges for temperature cycling test, Method 1010.8, Mil-Std-883H

Test condition	Cold	Hot
Dwell at each temperature extreme ≥ 10 min	With load temperature tolerances during recovery time	With load temperature tolerances during recovery time
A	−55°C; +0°C, − 10°C	+85°C; +10°C, −0°C
B	−55°C; +0°C, − 10°C	+125°C; +15°C, −0°C
C	−65°C; +0°C, − 10°C	+150°C; +15°C, −0°C
D	−65°C; +0°C, − 10°C	+200°C; +15°C, −0°C
E	−65°C; +0°C, − 10°C	+300°C; +15°C, −0°C
F	−65°C; +0°C, − 10°C	+175°C; +15°C, −0°C

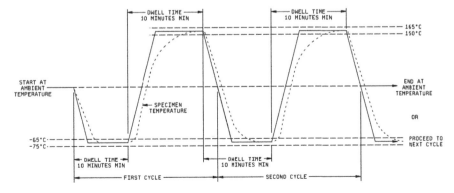

Fig. 6.22 Temperature cycling profile for Condition C. Mil-Std-883H, Method 1010.8 [23]

6.4.1.3 Variable Frequency Vibration, Method 2007.3

This vibration test is performed over a frequency range that is varied in an approximate logarithmic fashion between 20 and 2000 Hz and back to 20 Hz, with one cycle performed at 4 minutes minimum. Each orientation of the part, X, Y, and Z shall get 4 vibration cycles, for a total test time of 48 minutes minimum. Test conditions A, B, and C in Table 6.7 vary by peak acceleration. This test can be used for operational and non-operational conditions.

Table 6.7 Peak acceleration levels for Variable Frequency Vibration Test Conditions, Method 2007.3, Mil-Std-883H

Test condition	Peak acceleration (g)
A	20
B	50
C	70

6.4.1.4 Mechanical Shock, Method 2002.5

This mechanical shock testing method is for operational and non-operational conditions. This shock pulse is a half-sine condition. Peak shock levels and pulse durations are a function of test condition, shown in Table 6.8.

Table 6.8 Test conditions for mechanical shock, Method 2002.5, Mil-Std-883H

Test condition	Peak g level	Pulse duration in ms
A	500	1.0
B	1500	0.5
C	3000	0.3
D	5000	0.3
E	10,000	0.2
F	20,000	0.2
G	30,000	0.12

6.4.1.5 Constant Acceleration, Method 2001.3

This testing is performed for 1 minute in each of 6 axes, X_1, X_2, Y_1, Y_2, Z_1, Z_2. The Y_1 orientation is defined as the axis in which the device is separated from the mount in the package. Test condition E is typical, yet Conditions B, C, and D are often used (Table 6.9).

Table 6.9 Test conditions and g levels for constant acceleration, Method 2001.3. Mil-Std-883H

Test condition	Stress level in g
A	5000
B	10,000
C	15,000
D	20,000
E	30,000
F	50,000
G	75,000
H	100,000
J	125,000

6.4.2 Mil-Std-810 (Current Revision G)

This is a US Department of Defense standard for titled "Environmental Engineering Considerations and Laboratory Tests". Part One of this standard is the Environmental Engineering Program Guidelines and includes sections on General Program Guidelines as well as General Laboratory Test Method Guidelines. Part Two contains Laboratory Test Methods. Part Three is a Guidance section on World Climactic Regions, which is very helpful information for any reliability program. This standard contains Testing Methods on military-specific conditions such as gunfire shock, pyroshock, sand and dust, icing/freezing rain, freeze/thaw, mechanical vibrations of ship-board equipment, vibro-acoustic/temperature, and explosive atmosphere. Even if the part condition is not military yet does have to operate in extreme environments, some of these tests can provide guidelines for testing

Table 6.10 Two conditions for hot locations in the world based on diurnal cycles covered in Tables 6.11 and 6.12, Mil-Std-810G, Method 501.5 [24]

Design Type	Location	Ambient Air °C (°F)	Induced[a] °C (°F)
Basic Hot (A2)	Many parts of the world, extending outward from the hot dry category of the southwestern United States, northwestern Mexico, central and western Australia, Saharan Africa, South America, southern Spain, and southwest and south central Asia	30–43 (86–110)	30–63 (86–145)
Hot Dry (A1)	Southwest and south central Asia, southwestern United States, Saharan Africa, central and western Australia, and northwestern Mexico	32–49 (90–120)	33–71 (91–160)

[a] Induced conditions for extreme storage or transit environments

methodologies. This is a 804 page document so only a few test conditions are covered here. Test methods covered from Mil-Std-810 are high temperature testing and temperature shock.

6.4.2.1 High Temperature, Method 501.5

In this method, the reliability engineer must determine the proper testing temperatures through reading this extensive method. Guidelines on temperature exposure are based on actual elevated temperature conditions on the planet. Tables 6.10, 6.11 and 6.12 include hot conditions for diurnal cycles (24 h periods).

Table 6.11 Basic Hot (A2 from Table 6.10) diurnal cycles for ambient and induced conditions. Humidity values are included but not necessary to add for testing [24]. Mil-Std-810G, Method 501.5

| Time of Day | Ambient air conditions | | Induced (storage and transit) conditions | |
	Temperature °C (°F)	Humidity % RH	Temperature °C (°F)	Humidity % RH
0100	33 (91)	36	33 (91)	36
0200	32 (90)	38	32 (90)	38
0300	32 (90)	41	32 (90)	41
0400	31 (88)	44	31 (88)	44
0500	30 (86)	44	30 (86)	44
0600	30 (86)	44	31 (88)	43
0700	31 (88)	41	34 (93)	32
0800	34 (93)	34	38 (101)	30
0900	37 (99)	29	42 (107)	23
1000	39 (102)	24	45 (113)	17
1100	41 (106)	21	51 (124)	17
1200	42 (107)	18	57 (134)	8
1300	43 (109)	16	61 (142)	6
1400	43 (110)	15	63 (145)	6
1500	43 (110)	14	63 (145)	5
1600	43 (110)	14	62 (144)	6
1700	43 (109)	14	60 (140)	6
1800	42 (107)	15	57 (134)	6
1900	40 (104)	17	50 (122)	10
2000	38 (100)	20	44 (111)	14
2100	36 (97)	22	38 (101)	19
2200	35 (95)	25	35 (95)	25
2300	34 (93)	28	34 (93)	28
2400	33 (91)	33	33 (91)	33

Table 6.12 Hot dry (A1 from Table 6.10) diurnal cycles for ambient and induced conditions. Humidity values are included but not necessary to add for testing [24]. Mil-Std-810G, Method 501.5

Time of day	Ambient air conditions		Induced (storage and transit) conditions	
	Temperature °C (°F)	Humidity % RH	Temperature °C (°F)	Humidity % RH
0100	35 (95)	6	35 (95)	6
0200	34 (94)	7	34 (94)	7
0300	34 (93)	7	34 (94)	7
0400	33 (92)	8	33 (92)	7
0500	33 (91)	8	33 (92)	7
0600	32 (90)	8	33 (91)	7
0700	33 (91)	8	36 (97)	5
0800	35 (95)	6	40 (104)	4
0900	38 (101)	6	44 (111)	4
1000	41 (106)	5	51 (124)	3
1100	43 (110)	4	56 (133)	2
1200	44 (112)	4	63 (145)	2
1300	47 (116)	3	69 (156)	1
1400	48 (118)	3	70 (158)	1
1500	48 (119)	3	71 (160)	1
1600	49 (120)	3	70 (158)	1
1700	48 (119)	3	67 (153)	1
1800	48 (118)	3	63 (145)	2
1900	46 (114)	3	55 (131)	2
2000	42 (108)	4	48 (118)	3
2100	41 (105)	5	41 (105)	5
2200	39 (102)	6	39 (103)	6
2300	38 (100)	6	37 (99)	6
2400	37 (98)	6	35 (95)	6

6.4.2.2 Thermal Shock, Test Method 503.5

This test method uses a dual chamber system (Fig. 6.18), and is based on extreme temperature changes over a short period of time that the part will experience. There are many cycling conditions that are included in this test method; the example of multi-cycle air/air shocks is in Fig. 6.23. Here, the cycling starts and stops at ambient temperature, while high and low temperatures (T1 and T2) are based on conditions experienced as well as the Methods in Mil-Std-810G that provide information on extreme temperature ranges. The transfer rate is rapid in this test, 1 minute maximum between T1 and T2. The method allows for single shocks and multi-shock cycling.

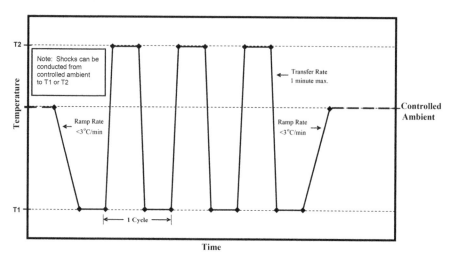

Fig. 6.23 Thermal Shock Multi-Cycling from Mil-Std-810G, Method 503.5 [24]

6.4.3 Telcordia Standards

There are many Telcordia standards that set reliability qualification testing conditions. Telcordia standards are written for telecommunication equipment, systems and services. GR-63-CORE is a set of documents for physical and environmental testing and criteria. Shock levels will be included from documents that are part of the GR-63-CORE family (Table 6.13) while high and low temperature storage and temperature cycling requirements are in Table 6.14. Note that GR-1073-CORE "Generic Requirements for Singlemode Fiber Optic Switches" was written for devices types including MEMS. GR-1221 is titled "Generic Reliability Assurance Requirements for Passive Optical Components". Optical switching is the RFMEA case study in Chapter 5.

6.4.4 Automotive Standards

AEC-Q100, Q101 and Q102 contain qualification standards and testing methods for the automotive industry. These are available free of charge. This series of specifications was written by the Automotive Electrics Council. AEC-Q100 grades the operational temperature ranges in the automotive environment, in Table 6.15. Qualification tests conditions in the AEC-Q10X series are a function of the Grade.

ESD testing is covered very nicely in this automotive standard. Human Body Model testing is covered in AEC-Q101-001A Human Body Model (HBM) Electrostatic Discharge Test (ESD). The Machine Model (MM) Electrostatic

Table 6.13 Set of mechanical shock and drop tests from standards in the GR-63 CORE family [25]. Mechanical shock requirements are based on the half sine pulse in a similar duration range as Mil-Std-883. Reference [25] reprinted with permission. Copyright 2007 SPIE

Requirement	Shock level	Duration	Number of shocks
GR-1073-CORE, environmental and mechanical criteria (component)	500G	1 ms pulse, half sine	2 shocks/direction/axis, 3 axes (12 shocks total)
GR-1221-CORE, reliability tests	500G	1 ms pulse, half sine	5 shocks/direction/axis, 3 axes (30 shocks total)
EIA/TIA-455-2A, drop testing; light service applications	Drop height 1.8 m (very high G levels)	Mount rigidly so shock is transmitted to internal components	8 drops/3 axes; 5 repetitions of entire cycle

Table 6.14 Thermal cycling, high temperature and low temperature testing conditions from GR-1073-CORE and GR-1221-CORE. Reference [25] reprinted with permission. Copyright 2007 SPIE

Requirement	Thermal stress	Duration	Paragraph
GR-1221-CORE, reliability tests	High temp storage, 85°C or max storage temp	2000 h	Table 4.2 endurance tests
GR-1221-CORE, reliability tests	Low temp storage, −40°C or min storage temp	2000 h	Table 4.2 endurance tests
GR-1221-CORE, reliability tests	Temperature cycling, −40°C to 70°C	100 cycles pass/fail, 500 cycles for information	Table 4.2, controlled environment application: endurance tests
GR-1221-CORE, reliability tests	Temperature cycling, −40°C to 85°C	500 cycles pass/fail, 1000 cycles for information	Table 4.2, uncontrolled environment application: endurance tests
GR-1073-CORE, environmental criteria	Temperature cycling, −40°C to 70°C	10 cycles	Table 5.1, Transport, storage and handling tests

Table 6.15 Grading system in Automotive Standard AEC-Q100 (current revision G) [26]

Grade	Automotive ambient temperature range
Grade 0	−40°C to +150°C
Grade 1	−40°C to +125°C
Grade 2	−40°C to +105°C
Grade 3	−40°C to +85°C
Grade 4	0°C to +70°C

Table 6.16 Three ESD Models and associated specifications

ESD Model	AEC-Q101 Standard	JEDEC Standard	ESD/EOS Standard
HBM	AEC-Q101-001A	EIA/JESD22-A114	ESD/EOS Association Specification STM5.1
MM	AEC-Q101-003E	EIA/JESD22-A115	ESD/EOS Association Specification S5.2
CDM	AEC-Q101-005	EIA/JESD22-C101	ESD/EOS Association Specification STM5.3.1

Discharge Test (ESD) is covered in AEC-Q101-003E. Charged Device Model (CDM) is detailed in AEC-Q101-005. These are reference ESD specifications by JEDEC and the ESD/EOS association. Table 6.16 organizes these standards by ESD model.

6.5 MEMS Qualification Testing

Automotive applications for MEMS inertial sensors were one of the first product applications for MEMS (Chapter 2, Section 2.5.2). Pressure sensors are also becoming a requirement for tire pressure in automotive applications. These two high volume MEMS devices are covered in this Sections 6.5.1 and 6.5.2 of MEMS Automotive Qualification Testing by reviewing early studies on reliability. Space and military examples are in 6.5.3 and 6.5.4, respectively.

6.5.1 ADI Accelerometers for Airbag Deployment

Early accelerometer designs and packaging types (for both high and low G devices) were tested to a series of qualification tests. Figure 6.24 shows these early designs while Table 6.17 gives design information for accelerometers A, B, and C. A total of 4590 devices were tested to the test methods in Table 6.18, based on Mil-Std-883 and AEC-Q100.

Testing methods included High Temperature Operating Life (HTOL) from Mil-Std-883, Thermal Shock from Mil-Std-883, Temperature Cycling from Mil-Std-883, High Temperature Storage from Mil-Std-883, Group D Subgroup 4 from

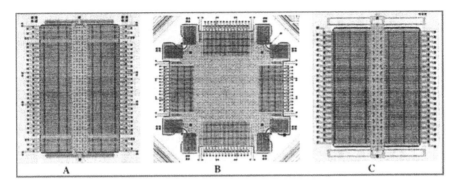

Fig. 6.24 Three early Analog Devices accelerometer designs. Reference [27] reprinted with permission. Copyright 1999 SPIE

Table 6.17 Device characteristics of early accelerometer designs A, B, and C. Reference [27] reprinted with permission. Copyright 1999 SPIE

Accelerometer design	A	B	C
Rated sensitivity range (g)	50 g	2 g	50 g
Axes of sensitivity	X	X, Y	X
Resonant frequency (in kHz)	24.5	10	25.5
Spring constant (X, in N/m)	5.4	3.0	4.0
Package	14-pin Cerpak	14-pin Cerpak	8-pin Cerdip

Mil-Std-883 which consisted of this sequence of testing performed on the same set of parts with electrical testing performed after each stress test.

- Mil-Std-883 Method 2002, Condition B, mechanical shock of 1500 g with 0.5 ms pulse width
- Mil-Std-883 Method 2007, Condition A, variable frequency vibration
- Mil-Std-883, Method 2001, Condition E, constant acceleration

Mechanical Drop was performed from a height of 0.3 meters onto a granite surface, in the X axis, Y axis and Z axis. Electrical test is performed before this is repeated at 1.2 meters. Random Drop is performed at 1.2 meters, for 10 drops, with test in between each drop. The sample sizes and test methods are summarized in Table 6.18.

Of the 4590 test samples, there were seven failures (Table 6.19); they were all attributed to mechanical failure mechanisms. Accelerometer B experienced a stiction failure yet this was during high temperature operating life testing. As that testing does not include mechanical shocks, it was thought the stiction failure experienced an unusually high mechanical shock in transport between test chamber and electrical testing. Accelerometer A had a particle failure. Elemental analysis and source identification for elimination is recommended for contamination type failures

Table 6.18 Testing to qualification standards performed on early Analog Devices accelerometer designs. Reference [27] reprinted with permission. Copyright 1999 SPIE

Stress	Conditions	Device type	Quantity per lot	Number of lots	Total devices
HTOL	Mil-Std-883	A	45	3	135
	Method 1005	B	45	3	135
	Condition C	C	45	3	135
Thermal Shock	Mil-Std-883	A	45	3	135
	Method 1011	B	135	3	405
	Condition C	C	45	3	135
Temperature	Mil-Std-883	A	45	3	135
Cycle	Method 1010	C	45	3	135
	Condition C				
High	Mil-Std-883	A	135	3	405
Temperature	Method 1008	B	180	3	540
Storage	Condition C	C	135	3	405
Group D	Mil-Std-883	A	45	3	135
Subgroup 4	Method 5005	C	45	3	135
Mechanical Drop	0.3 m X,Y,Z;	A	45	3	135
	1.2 m,X,Y,Z	C	45	3	135
Random Drop	1.2 m, 10 drops	A	135	3	405
		B	180	3	540
		C	135	3	405

Table 6.19 Seven failures out of 4590 devices were observed and failure analyzed. Reference [27] reprinted with permission. Copyright 1999 SPIE

Device type	Stress test	Failure mechanism
Accelerometer B	HTOL	Stiction
Accelerometer A	Random Drop	Particle Impedance
Accelerometer C	Random Drop	Jump Shift
Accelerometer C	Random Drop	Jump Shift
Accelerometer C	Random Drop	Jump Shift
Accelerometer C	Mechanical Drop	Jump Shift
Accelerometer C	Mechanical Drop	Jump Shift

(Chapter 3). The most interesting part of this study was the failures of Accelerometer C. Figure 6.25 shows the jump shift failure.

This was a very early spring design. Although the failure rate in the field was just 0.218 ppm [27] for jump shifting, this indicates that although random and mechanical drop tests are very severe, they can bring out mechanical failure at high acceleration factors.

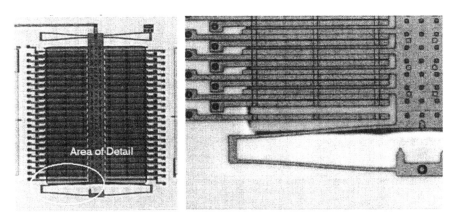

Fig. 6.25 Early spring design showed jump shift failure. Reference [27] reprinted with permission. Copyright 1999 SPIE

6.5.2 *Motorola MEMS Pressure Sensors*

Motorola performed studies on their early MEMS packaging design for infiltration of various liquid media that could result in failure. In the automotive application of tire pressure monitoring, of this type of testing is very important [28, 29]. In addition, acceleration testing was performed on the Motorola device to determine output changes as a function of tire rotation. A review of these studies by the Motorola engineers follows. These examples show how the Motorola MEMS engineers were clever in developing tests to stress their design for flaws and simulate operational acceleration. Figure 6.26 illustrates the early package design with Fluorogel encapsulant.

This package was subjected to the media and conditions in Table 6.20. Media, ingredient, pH, temperature, pressure of environment, and testing apparatus are covered. Devices were powered under testing with failure criteria of zero level capacitance shift $\leq 25\%$, sensitivity shift $\leq 2.5\%$, and impedance shift $\geq 2.5 \times 10^6$.

Fig. 6.26 Early Motorola pressure sensor package. Reference [28] reprinted with permission. Copyright 1999 SPIE

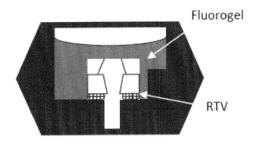

Table 6.20 Media testing matrix for Motorola pressure sensors. Reference [28] reprinted with permission. Copyright 1999 SPIE

Property	Purpose	Ingredients	pH	Temperature	Pressure	Testing Apparatus
Nitric acid	Strongly acid media	Strong acid, fully dissociated in water	1.8	85°C	=2.5 in. of water	Individual exposure modules
Seawater	Neutral media	Sodium chloride, sulfate, magnesium, calcium, potassium	8.4	65°C	Atmospheric	Caustic exposure tank
Organic/aqueous solution	Organic media	80–90% water, 1–10% ethylene glycol, 10–20% IPA, 1–10% alkyl olefin sulfonate	–	65°C	Atmospheric	Caustic media exposure tank

Table 6.21 Percentage of passing parts by interval and media test. Reference [28] reprinted with permission. Copyright 1999 SPIE

Media	Interval		
Test	A	B	C
Nitric acid	60	26	26
Seawater	22	6	6
Organic/Aqueous solution	39	6	0

Table 6.22 Interval times for the various media tests. Reference [28] reprinted with permission. Copyright 1999 SPIE

Removal Time (h)	Organic solvent	Seawater	Nitric Acid
Interval A	162	163	158
Interval B	280	347	318
Interval C	500	515	504

Results from this testing are shown in Table 6.21 as percent of passing devices tested, shown as percentage of total samples. Three sequential time intervals were tested in an ex-situ fashion (Table 6.22). Interestingly, the nitric acid testing was the least aggressive. Seawater was thought to be more aggressive due to chlorine content and corrosion, and the organic/aqueous solution can swell and chemically change the protective encapsulant.

To add theory to this media study, an electrochemical test method was developed to test the Motorola pressure sensor using impedance spectroscopy and alternate encapsulants [29], with open circuit potential and polarization used to test for adhesion strength and permeability of the encapsulation (see Chapter 4).

Learnings:

- What chemistry (media) is most destructive
- Encapsulant interaction with liquid media
- Kinetics of the corrosion mechanism
- Relationships between corrosion measurements and failure mechanisms

Acceleration sensitivity was measured using a constant acceleration technique for early Motorola pressure sensors. During tire pressure monitoring, the pressure sensor voltage output will change as a function of the acceleration experienced. In this case constant acceleration testing is a simulation of the operating environment, and not just a check on design and materials of construction (Section 6.4.1.5). Figure 6.27 is the constant acceleration set up consisting of a rotating disk, slip-ring

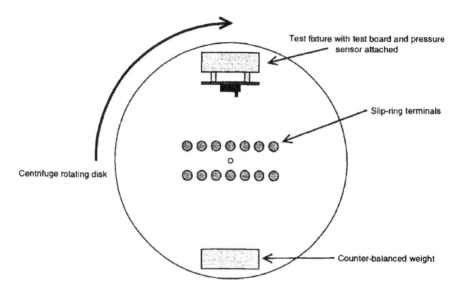

Fig. 6.27 Illustration of constant acceleration set-up for pressure sensor testing. Reference [30] reprinted with permission. Copyright 1999 SPIE

terminals, test fixture for insitu output monitoring during testing, and a counter-balanced weight. Testing was performed at three constant acceleration levels (that simulated high vehicle speed) to detect changes in pressure sensor output (Fig. 6.28). This methodology can be applied to pressure sensors of various geometries to determine change in output due to tire rotation during automotive pressure monitoring.

6.5.3 Example: Space and Military Qualification

A NASA level random vibration test is discussed, and test results of a MEMS DMD product ruggedized into a display system for military applications are presented. Space radiation testing is detailed in the review section in Section 4.4.1.

6.5.3.1 NASA Space Random Vibration Specifications

Table 6.23 is a summary of the random vibration specification determined by thorough analysis of launch vibration conditions. Sources of vibration include acoustic vibrations, engine firing, and turbo pumps, etc. These conditions are specific to spacecraft design and assembly-level mounting, etc. The methodology to determine these specifications and many other space test conditions are outlined in reference [31]. Although these conditions are design specific, it is interesting to see that the 20–2000 Hz frequency range is the same as Mil-Std-883H.

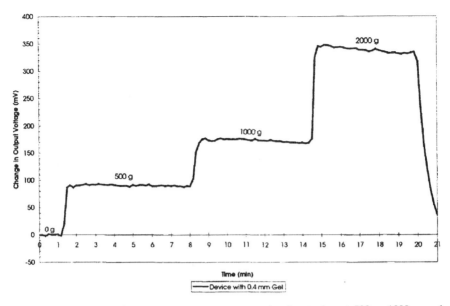

Fig. 6.28 Early Motorola pressure sensor constant acceleration testing at 500 g, 1000 g, and 2000 g levels, versus change in output voltage. Reference [30] reprinted with permission. Copyright 1999 SPIE

Table 6.23 NASA reliability random vibration specification for flight acceptance test. Reference [31] Courtesy NASA/JPL-Caltech

Spacecraft-level		Assembly-level	
Frequency (Hz)	Level	Frequency (Hz)	Level
20–45	+10 dB/octave	20–80	+6 dB/octave
45–600	0.06 g^2/Hz	80–1000	0.25 g^2/Hz
600–2000	6 dB/octave	1000–2000	−12 dB/octave
Overall	**7.7 G(rms)**	**Overall**	**17.6 G(rms)**

Duration:
DESIGN: 3 MINUTES IN EACH OF 3 ORTHOGONAL AXES
PF TEST: 2 MINUTES IN EACH OF 3 ORTHOGONAL AXES
ACCEPTANCE SAME AS PF

6.5.3.2 DMD Ruggedization – Example of a MEMS Product Tested to Extreme Conditions

An interesting reliability study was performed on a display product that was ruggedized for military applications and was called the Raytheon 1210 Digital Ruggedized Display (1210 DRD), however, a MEMS 1280 × 1024 DMD (Chapter 2) was part of the system. A Highly Accelerated Life Test (HALT) system manufactured by the Qualmark® Corporation was used as a test chamber. This environmental test platform has the ability to test for cold limit, hot limit, thermal shock testing, vibration, as well as the combination of thermal and vibration testing. As this testing is extreme, reliability lifetime predictions cannot be made, but the testing itself brings out design weaknesses and allows for root cause analysis (Chapter 5) to be performed on failures during testing. The 1210 DRD testing was design to meet the Common Large Area Display Set (CLADS) Program [32]. The HALT testing was performed and compared to the CLADS system requirements. Table 6.24 summarizes the temperature and vibration testing results.

Expanded and contracted images were attributed to athermalization failure, yet the flicker failure was not detailed. As HALT testing uses the Test, Analyze and Fix (TAAF) method where failures are recognized, failure analyzed and fixed, and the testing continues. In this way, the corrective action is tested quickly and with the same test conditions that created earlier failure. Table 6.25 shows conditions of simultaneous vibration/thermal cycling testing. With the exception of oscillator failures on printed circuit boards in cycles 6 and 8, no additional failure occurred.

This example shows how a MEMS DMD performed well at extreme testing in a ruggedized display designed for military applications.

Table 6.24   Results of 1210 DRD extreme HALT testing [33]

Testing	CLADS specification	1210 DRD performance in HALT	Comments
Cold Limit	Operational from −20°C	Operated to −71°C	Video image expanded but worked at −50°C, flickered at −71°C; continued to perform
Hot Limit	Operational to 55°C	Operated to 97°C	Color wheel tape failed at 65°C, no damage to MEMS DMD; at 55°C video image contracted but worked, lamp went out at 97°C but started working again when temperature reduced to 80°C
Thermal Shock	−20°C to 55°C, 15°C per minute ramp rate	−80°C to 100°C, 60°C per minute ramp rate	18 cycles performed in HALT, system worked but screen delaminated, MEMS had no damage
Vibration	Operational: 2.8 g rms	Tested up to 25 g rms	No failures
Vibration	Endurance: 5.9 g rms	Tested up to 25 g rms	No failures

Table 6.25 Combined vibration and thermal cycling testing of 1210 DRD. Reference [33] reprinted with permission. Copyright 1998 SPIE

Cycle	Temperature range	Vibration level
Cycle 1	−35°C to +70°C	15 G (RMS)
Cycle 2	−45°C to +80°C	20 G (RMS)
Cycle 3	−55°C to +90°C	25 G (RMS)
Cycle 4	−65°C to +95°C	30 G (RMS)
Cycle 5	−70°C to +95°C	30 G (RMS)
Cycles 6–10	−70°C to +95°C	30 G (RMS)

6.6 Summary

MEMS testing and qualification are scientific areas where creativity and current equipment and/or test methods can be combined to produce accurate production and reliability testing, as well as testing for qualification. Examples of unique test methods for MEMS are included. Reinventing the wheel isn't required, but use of new methods based on existing equipment or adding equipment to existing test platforms is common. Qualification standards that exist today are covered with examples of hard to find test data on early Analog Devices accelerometers, Motorola pressure sensors, and the MEMS DMD for ruggedized display application.

References

1. Douglass, M.R. (2008) MEMS Reliability, Coming of Age. Proc. of SPIE. 6884, 688402, (eds. Hartzell, A.L. Ramesham, R.).
2. Zunino, J. III, Skelton, D. (2006) Department of Defense Need for a Micro-electromechanical Systems (MEMS) Reliability Assessment Program. Proc. of SPIE. 5716, (eds. Tanner, D., Rajeshuni Ramesham R.).
3. Merlijn van Spengen, W. (2003) MEMS reliability from a failure mechanisms prospective. Microelectronics Reliability 43, 1049–1060.
4. Tanner, D., et al. (2000) MEMS reliability in a vibration environment. IEEE IRPS, San Jose, CA, April 10–13, pp. 139–145.
5. http://www.bostonmicromachines.com/news_press_ao_toolkit.htm.
6. Dagel, D.J., et al. (2006) Large-stroke MEMS deformable mirrors for adaptive optics. J. Microelectromech. Syst. 15(3).
7. Andrews, J.R., et al. (2008) Performance of a MEMS Reflective Wavefront Sensor. Proc. of SPIE Vol. 6888, 68880C, MEMS Adaptive Optics II.
8. Zecchino, M., Novak E. (2003) MEMS in Motion, a New Method for Dynamic MEMS Metrology", Veeco Instruments, AN514-1-0603.
9. Veeco Instruments, Wyko NT Series Optical Profilometers. B506, Rev A7, 2009.
10. http://www.veeco.com/Products/metrology_and_instrumentation/Optical_Profilers/Wyko_DMEMS_NT3300/index.aspx.
11. Cascade Microtech and Polytec PI. MEMS Optical Switch Production Test System. document number MEMSPB-0401.
12. Courtesy W. Noell at the EPFL and S. Weber at the University of Geneva.

13. Lawrence, E. Rembe, C. (2004) MEMS characterization using new hybrid laser doppler vibrometer/strobe video system. Proc. of SPIE. 5343, Reliability, Testing and Characterization of MEMS/MOEMS III.
14. Tanner, D. et al. (1997) First reliability test of a surface micromachined microengine using SHiMMer. SPIE, 3224.
15. http://www.cszindustrial.com/products/tempchambers/ctseries.htm.
16. http://www.cszindustrial.com/products/microclimate/microclimate.htm.
17. JESD22-A104C; JEDEC Standard, Temperature Cycling, May 2005. JEDEC Solid State Technology Association.
18. http://www.cszindustrial.com/products/thermalshock/vtsseries.htm.
19. http://www.cszindustrial.com/products/thermalshock/tsbseries.htm.
20. O'Reilly, R. (2006) High G testing of MEMS devices. Proc. of the SPIE. 6111, Reliability, Packaging, Testing, and Characterization of MEMS/MOEMS V".
21. Seinfeld, J. (1986) *Atmospheric chemistry and physics of air pollution*, New York: John Wiley and Sons, p. 181.
22. Hartzell, A.L., et al. (2010) Reliability of MEMS deformable mirror technology used in adaptive optics imaging systems. Proc of SPIE. 7595, 75950B. MEMS Adaptive Optics IV.
23. Mil-Std-883H.
24. Mil-Std-810G.
25. Bhattacharya, S., Hartzell, A. (2007) J. Micro/Nanolith, MEMS MOEMS, Jul-Sep /Vol 6(3).
26. AEC-Q100G.
27. Delak, K., et al. (1999) Analysis of manufacturing scale MEMS reliability testing. SPIE Vol. 3880, Part of the SPIE Conference on MEMS Reliability for Critical and Space Applications, Santa Clara, CA.
28. Gogoi, B. (1999) et al. Integration Issues for Pressure Sensors. Proc. SPIE, 3874, 174.
29. Bitko, G. et al. Analytical techniques for examining reliability and failure mechanisms of barrier coating encapsulated silicon pressure sensors exposed to harsh media. Proc. SPIE, Vol. 2882.
30. August, R., et al. (1999) Acceleration sensitivity of micromachined pressure sensors. SPIE Vol. 3876. Part of the SPIE Conference on Micromachined Devices and Components.
31. Stark, B. (ed) (1999) MEMS reliability assurance guidelines for space applications. JPL. Publication 99-1, National Aeronautics and Space Administration, Jet Propulsion Laboratory, California Institute of Technology, Pasadena, CA. January.
32. Spiegl, C. (1997) Large-area display system using Digital Light Processing. 1997 SPIE International Symposium on Aerospace/Defense Sensing and Controls, Cockpit Displays IV, Orlando, FA.
33. Becker, B., Phillips, R. (1998) Highly accelerated life testing for the 1210 Digital Ruggedized Display. Proceedings of the SPIE Vol. 3363. Cockpit Displays V: Displays for Defense Applications, 1998.

Chapter 7
Continuous Improvement: Tools and Techniques for Reliability Improvement

One common feature of MEMs enabled products that have crossed the threshold of prototype volumes into large-scale volume production is that in a majority of cases, the product development effort from initial prototype to final market insertion lasted much longer than planned. A major cause of this has been designing in product reliability which has tended to be more of an afterthought rather than part of an active engineering effort at the start of the development cycle. In Chapter 1, we mentioned that a significant time-consuming factor has been the persistence of a "traditional" manufacturing approach, with repeated iterations to develop higher reliability in the product, which leads to much longer development times [1, 2]. Many factors such as the novelty of MEMS technology, lack of adequate design tools, reliability, and loosely connected engineering of the MEMS device, ASIC & custom packaging have contributed to this methodology. In addition, we have seen that every MEMS device has a unique fabrication process flow and unlike semiconductors there is no "standard" process, which makes it challenging to collect accurate material properties, limits understanding of processing effects on materials, and process variability.

A more proactive approach (shown in Fig. 1.5[1]), which includes reliability considerations at the very beginning of the development effort rather than towards the end is definitely necessary to improve product reliability and time-to-market. This implies focus on developing reliability predictions and mitigating potential failure modes more effectively through design or manufacturing. As discussed in earlier chapters, the process of identifying failure modes and estimating their potential effects is accomplished through an FMEA analysis.

The primary benefit of employing such a strategy for MEMS product development is faster design cycles [3], a faster path to volume manufacture and most importantly, improved yield and reliability. In this chapter, the main focus will be on techniques for improvement of both yield and reliability for MEMS products but first we need to discuss the connection between these important terms.

---

[1]Chapter 1

A.L. Hartzell et al., *MEMS Reliability*, MEMS Reference Shelf,
DOI 10.1007/978-1-4419-6018-4_7, © Springer Science+Business Media, LLC 2011

7.1 The Yield-Reliability Connection

Yield in manufacturing is defined *as the number of products that can be sold divided by the number of products that can be potentially made.* In the MEMS industry, yield is represented by the functionality and reliability of sensor devices produced on the wafer surfaces and is measured as the percentage of chips in a finished lot[2] that pass all tests and function as specified. As described in Chapter 1, reliability is defined as a *probability that a component part will satisfactorily perform its intended function for a specified period of time.* In basic terms, a product with superior reliability is perceived by the customer to be of superior quality, but to achieve this improved reliability there is a need to incorporate a few essential components such as:

- Qualified Manufacturing Processes
- Design for Reliability (DfR)
- Failure Mode prevention

In the MEMS industry, this is not as straightforward or simple for a few reasons:

- MEMS fabrication processes usually require custom development which is time consuming and needs multiple wafer starts to debug process issues. The maturity of the manufacturing process is directly correlated to the yields achievable. Average MEMS fabrication process yields in the 90% range are not uncommon these days but they are not as high as semiconductor ICs.
- Design Tools to simulate and predict reliability are continuously improving but the lack of accurate material behavioral models makes it challenging to simulate potential failure modes, and finally
- FMEA analysis during the development phase is a systematic technique to identify potential failure modes that can help minimize field failures but as with the development of new process technologies, it can be quite challenging to capture all potential failure modes.

In semiconductor manufacturing, the *Six Sigma*[TM][3] methodology is widely accepted as a way to improve quality and achieve continuous improvement, and this methodology [4] is also employed within the MEMS industry. *Six Sigma*[TM] works to identify and remove the causes of defects as well as to minimize variability in all manufacturing process steps. The methodology involves the creation of a specialized infrastructure within the organization and uses statistical methods to achieve quantified process related improvements or financial targets. Basically,

---

[2]A manufacturing *lot* can be defined rather arbitrarily as all the dice on a single wafer, or an assembly lot which could be an arbitrarily defined number of packaged parts.

[3]*Six Sigma* is a Motorola trademark – "Six Sigma can be seen as: a vision; a philosophy; a symbol; a metric; a goal; a methodology" – [4].

Six Sigma or 6σ implies that a given manufacturing process has six standard deviations (6σ) from the *mean* of the process (which is assumed to be a distribution[4]) to the nearest specification limit, and this results in very few parts that would fail to meet the specification. In most real-world cases, 99% or 4σ indicates a product or service of sufficiently high quality, but in some cases 4σ quality can be seen to be inadequate. For example, 4σ quality means 20,000 lost pieces of mail every hour or unsafe drinking water almost 15 min every day or 5000 incorrect surgical operations per week, or 2 short or long plane landings each day. Clearly, it is apparent that this level of quality is not sufficient for applications such as semiconductor chips or MEMS sensors, where the volumes of components manufactured could be in 10 s of millions, and the expectation of quality is measured in PPM (parts-per-million).[5]

In Fig. 7.1, the ±3σ area under the curve of a normal distribution is 99.73% of the total area under the curve, and this corresponds to parts that are within the specification limits and 0.27% of parts are out of specification which is 2700 PPM. In manufacturing processes, statisticians have found that processes often shift over time and that shift can be as high as 1.5σ from mean [4]. This implies that a process that starts off as 6σ could end up being a 4.5σ process in the future. When this is applied to a 3σ process, a shift of 1.5σ from the center of the distribution means that only 93.32% of the parts meet specification corresponding to a 67,000 PPM failure rate which is of very poor quality.

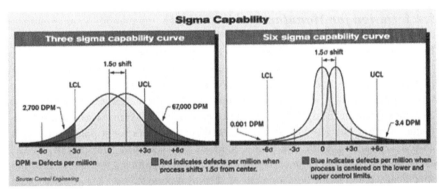

Fig. 7.1 Achieving product and process quality – three sigma (*left*) and six sigma (*right*) (reprinted with permission copyright - Control Engineering, www.controleng.com, CFE Media)

Another often used reliability measure is the process indices which are also called the *capability indices* (c_p & c_{pk}). The c_p measures the variability of the process and is defined as the *ratio of the maximum allowable range of a characteristic to the normal ±3σ variation*, while c_{pk} measures how close the process is running to the specification limits.[6] As the process standard deviation goes up or the process mean

---

[4]See Chapter 2 Sec. 2.3

[5]10,000 ppm = 10,000 parts/million = 1%.

[6] $c_p = (USL–LSL)/6\sigma$; $c_{pk} = Min[(USL–mean)/6\sigma, Mean–LSL)/6\sigma]$.

moves away from center, the number of standard deviations that will fit between the mean and the nearest limit decreases. The c_p does not take account of how well the process distribution is centered within the limits and the c_{pk} index is usually employed. For example, a process that operates with 6σ will have a c_p and c_{pk} = 2 and a failure rate of less than 2 parts per billion and is capable of producing extremely reliable products or services.

In the next subsection we will look at techniques to improve quality both from a manufacturing yield and reliability perspective. Yield improvement is a critical task in any product development effort although there are instances where a product will be released into the market before it has optimized processes in manufacturing.

7.2 Yield Improvement Techniques

In this section, we will discuss specific techniques to improve yield and thus improve overall quality. The main topics covered in this section are design for manufacturability (DfM), design-for-test (DfT), sensor- package integration and functional yield modeling for MEMS.

7.2.1 Design for Manufacturability (DfM)

In MEMS, incorporating DfM principles have been studied [2, 3] in some detail but we will highlight a few salient points in this section. DfM is a well known design philosophy and methodology that incorporates the manufacturing process as part of the design process [5], and broadly includes organizational changes, systematic concurrent engineering design principles and a common CAD framework for producing manufacturable designs with accurately predicted behavior for scalable, repeatable, and cost-effective volume production. Some of the basic DfM principles [6] are well established concurrent engineering practices but are significant because of how they specifically apply to the development of MEMS products:

- *Minimal number of components* – The use of fewer masks and fabrication steps, shortens development and increases manufacturability by simplification.
- *Use of Standard Modular Components* – In MEMS, components such as fabrication steps e.g. polysilicon deposition, or design components e.g. comb finger drives, with standard interfaces to reduce the research or engineering effort and that are capable of being assembled into more complex systems.
- *Designing parts for multi-use* – standard process flows for example could be used for multiple MEMS products. This has enormous potential to cut costs and save time by reducing the need for developing new process steps.
- *Design for ease of manufacturing* to reduce the amount of custom process development.

As we have seen in Chapter 3, there are several failures modes that can be traced directly to the design and manufacturing phases and minimizing the occurrence of these failures modes can have a positive effect on yield improvement and quality. In a typical product development flow, the product design phase has several overlapping phases [3] that flow together, and this is occasionally referred to as a *waterfall* model (Fig. 7.2).

Fig. 7.2 Waterfall model for product development

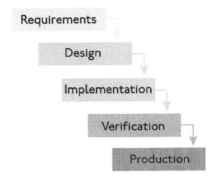

Technical Requirements: At this stage, the needs of the customer such as performance, cost target (ASP[7]), and footprint, are paramount to developing a technical specification for the product. In practice, the engineering team might maintain a customer requirements document which captures the original intent of the customer and which will also form the basis for developing the product data sheet once the product is launched. Once this has been achieved, there is a critical step to convert these requirements into engineering requirements and specifications which are required inputs for the design phase.

Conceptual Design Phase: A common and effective strategy to develop an optimized manufacturable design is to initially develop multiple design concepts [3]. These conceptual designs need to be rapidly evaluated and sorted based on top-level performance and manufacturing requirements. In IC design, this is achieved through parameterized *Component Libraries* and *Process Design Kits* that are now available to MEMS designers as well [7]. The growth of MEMS system level design tools and independent foundries around the globe has created a wide variety of design kits and MEMS component libraries [8, 9] that are now routinely used in the conceptual design phase. The MEMS component libraries [10] are basically parameterized behavioral models (similar to transistor models in IC design) in various coupled physics domains such as electro-mechanical or magneto-mechanical, and are available from several vendors (Coventor-*Architect*^{TM}, Intellisense – *Synpl*^{TM}, SoftMEMs – *MemsPro*^{TM} etc.). In a particular component element, the designer has access to all geometrical and material property variables, enabling easy exploration of process or design variables to evaluate a design or improve the robustness of the design.

---

[7] ASP – average sale price.

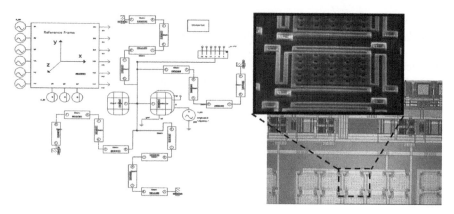

Fig. 7.3 Schematic Model of a MEMS variable capacitor realized in Architect^{TM} (reprinted with permission Copyright – 2004 SPIE [2])

The designer may use components from these libraries to create a model of the conceptual design of interest [11, 12], and the use the available network simulator (*Cadence*^{TM} or *Spice*® or *Saber*^{TM}) to predict performance under certain conditions. An example from [2] is reproduced in Fig. 7.3 above and shows a MEMS variable capacitor built within the system level simulator *Architect*^{TM} [13]. The capacitor comprises of several folded suspension beams or springs, a square mass or plate, and a gap capacitor, and these are built from components within the library of elements, such as beams, masses, coupled electro-mechanical elements parallel plates, and anchors. In individual elemental models, like for instance the structural beam element – a single beam element model is available of varying complexity i.e. linear or non-linear, 2nd order or higher, etc.) with access to geometrical parameters such as length (l), or thickness (t), or width (w), and material properties such as Young's modulus (E), or Poissons ratio (v), or residual stress and/or stress gradient. In some cases, a particular foundry may make a process design kit available which sets design constraints such as thickness or C_D but other design parameters like the geometry, may be varied. In this way the designer can create models of design concepts quickly, evaluate and filter designs based on a network level simulation in a standard network simulator.

The first obvious benefit is that the designer has the ability to be able to *set process constraints* which means that only those designs that can be manufactured within a particular process are evaluated. In Fig. 7.4, like the geometry, this ability to set process constraints is demonstrated using the rotation of a MEMS mirror as an example. The percentage perturbation of process and design variables on this performance metric allows the designer to, within a short time, tweak or discard a particular concept based solely on performance and manufacturability. The other major benefit is the ability to include the rest of the *system level* such as the control system electronics, or application level system description during the evaluation of the conceptual designs. In the case of the variable capacitor example, it is possible

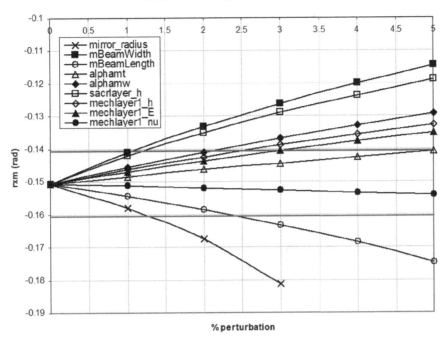

Fig. 7.4 Virtual manufacturing – sensitivity of mirror resonance frequency to mirror geometry (reprinted with permission – 2004 Coventor Inc. [13])

to include other circuit elements such as transistors, passives and the higher level system e.g. the variable capacitor could be analyzed as part of a MEMS based voltage-controlled oscillator (VCO). The reduction in complexity of the models, and the hierarchical nesting allows a wide variety of other performance metrics to be gathered about a conceptual design e.g. testability, by embedding the MEMS model in a virtual test environment. Following this approach, the design team can rapidly converge on a few designs that meet both performance and manufacturability requirements before proceeding to a more detailed design phase.

Detailed Design Phase: The key task in the detailed design phase is a *Design of Experiments* (DOE) to thoroughly analyze and optimize the performance of the conceptual design(s) selected in the previous phase, using a variety of standard semi-analytical, analytical and full-field numerical analysis techniques such as FEA/FEM, BEM[8] to perform *optimization* studies, *tolerance based design*, and *statistical design* to create a manufacturable part.

---

[8]BEM: boundary-element method.

The use of parameterized models described earlier is once again useful because once the design has been centered and critical process tolerances have been "cornered", the schematic can be automatically converted to a 2D (layout) or 3D solid model. The 3D solid CAD models are required starting points for physics based solvers to numerically simulate 3D field physics such as electrostatics, mechanics, coupled electromechanics, magnetics, electro-thermal effects, fluidic and fluid–structure-coupling effects (example of squeeze film damping analysis in an accelerometer – Fig. 7.5). It has always been quite challenging to simulate full device level coupled electro-mechanical MEMS design problems but with recent improvements in design tools [7] and increases in computing capacity these types of simulations and optimizations are routine and have contributed greatly to the reduction of design time and increased design verification.

Fig. 7.5 Coupled fluid-structure interaction simulation – squeeze-film damping in a mems accelerometer. Reprinted with permission of Comsol AB. Comsol and Comsol Multiphysics are trademarks of Comsol AB

Another critical design task that needs to be accomplished in the design phase is "virtual manufacturing", which basically includes the simulation of device performance over process tolerances and corners; and verification of manufacturability and functional yield in the presence of realistic process tolerances and testing. Two very useful analyses used are Monte Carlo[9] and sensitivity analyses – which are among several methods available for analyzing the effects of process tolerances and uncertainty. In this case, the designer is trying to determine how *random variations* and *errors* affect the *sensitivity*, *performance*, or even *reliability* of the system. Inputs may be randomly generated from *probability distributions* such as the thickness variation of a polysilicon deposit step or similar process tolerance. The data generated from the simulation is also represented as probability distributions or converted to *reliability predictions*, or *confidence intervals*. In [2], there are some practical examples that highlight the virtual manufacturing analysis of a MEMS variable capacitor and a bulk micromachined comb drive resonator.

In the design phase, it is necessary to the extent possible, to analyze the sources of manufacturing performance variation without actually fabricating actual devices.

---

[9]Class of computational algorithms employing random sampling – see [13]

In situations where fabrication processes are relatively mature (or fixed), actual process limits are available to predict functional yield of the device and the sensitivity of the design to process design rules; the use of such techniques can be quite advantageous but in cases where process steps are still evolving, it becomes necessary to quantify the process limits before attempting such detailed design. Obviously, this type of analysis is only possible when both the design and process engineering groups collaborate and share data applicable to the product i.e. process tolerances, and design rule violations [4].

Manufacturing Phase: The manufacturing phase begins simultaneously with design and essentially encompasses three primary activities – fabrication process, assembly process and final test:

- Develop MEMS Fabrication Process – There are two essential steps involved in developing the MEMS fabrication process – defining the process requirements and evaluating available processes.[10] In MEMS there are quite a few process technologies that have been demonstrated but the most common technology is surface micromachining, followed by bulk micromachining, LIGA etc. [14]. Within each of these technologies, there are multiple unit process steps that either deposit or remove material from the wafer surface, each with their own unique characteristics of surface chemistries and roughness, and geometry. In defining the process requirements, it is necessary to factor in the tolerance limits and variations of these critical process characteristics such that specific performance parameters of the MEMS device are constrained appropriately.

The evaluation of available fabrication processes becomes necessary because there are basically three paths for process selection with differing risks and cost structures – (a) available standard process – such as the MUMPS® process from MEMsCap, or (b) an available standard process with minor modifications, which could be in the form a different design rule or a thicker deposit etc., or even a well established process that is transferred to another facility, and (c) finally, a completely new process flow that is created from available unit process steps such as PECVD deposition and silicon fusion bonding etc., but which has never been used before to fabricate or produce a complete device. One obvious consideration is the ownership or control of the available process – most high volume manufacturing process flows only exist in captive fabs but that is changing and today, there are several foundries around the world that provide contract manufacturing services.[11] In Fig. 7.6 which has been adapted from [2], the pull-in voltage behavior of the variable capacitor element discussed earlier, is designed with the 3σ variation of the process. The resulting measurements show very good correlation between the process and the design.

---

[10]"Available processes" include standard processes steps and flows from external foundries.

[11] Available foundry services worldwide – http://www.yole.fr/pagesan/products/memsfoundries.asp

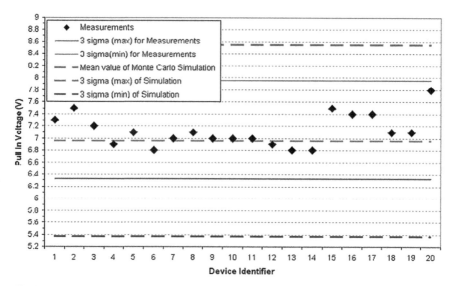

Fig. 7.6 Measured pull-in voltages for 20 suspended plate variable capacitors in comparison to simulation results (reprinted with permission Copyright 2004 SPIE [2])

- Packaging or Assembly Process – Preliminary assessment of the packaging options available and development of the assembly flow, bill-of-materials, and package design must be initiated simultaneously with the MEMS design. As we saw in Chapter 3, there are a number of failure modes related to the package and the selection of an appropriate assembly flow must address possible failure modes that impact production yield as well as reliability.

Today, plastic over-molded packaging [15] is quite common for most consumer MEMS products such as inertial sensors (accelerometers, and gyros) but in some cases specialty packaging is needed e.g. DMD® [1]. Further, decisions such as pad configuration, dimensions & material choice become important factors to minimize cost and improve overall yield. In MEMS, the interaction of the sensor and the package is a huge challenge and also has to be factored into the design process. It is often practical to perform the package design as an extension of the sensor design due to the coupling between the sensor and the package.

- Final Test Process – One manufacturing area that often gets overlooked during the design phase is the testability (further discussed in the next section) of the part. The sensor and package designers must concurrently engineer testability into the final part and this requires a detailed analysis of the part's electrical details such as the test modes i.e. switching the part into a mode other than its normal sensing operation so that certain types of parametric testing can be accomplished. A simple example of a test mode is a mode to measure the resonant frequency and Q factor of the sensor. There are mechanical details to address as well, such as the socket design or rate of testing (how many units per hour) among others that can influence the final design and manufacturing yield.

Design for Manufacturability (DfM) is a powerful methodology that both the design and manufacturing teams need to adopt to improve manufacturing yield in MEMS enabled products. The maximization of manufacturing yield based on device performance has a direct impact of increasing quality.

7.2.2 Design for Test (DfT)

Design for Test (DfT) is the collective name for techniques that add testability features to the part design. Automated and integrated test technology has become critically important for high-volume MEMS manufacturing and to improve reliability especially in safety critical applications such as automotive or other industrial applications. MEMS sensors *transduce* a non-electrical input such as acceleration or pressure or light, into an electrical time-varying output and present quite a different challenge for high-volume automated testing compared to traditional ICs [16]. Figure 7.7 shows a typical high volume test platform used for MEMS product testing. In the development of a similar high volume test solution for MEMS parts, it is necessary to be able to measure parameters like current consumption, bandwidth, self-test (to ensure operation) and other parameters (typically important to the factory i.e. resonant frequency, noise etc.) in production because without direct measurement of these parameters it is impossible to improve yield.

Fig. 7.7 High volume automated MEMS test equipment (Copyright Teradyne Inc., Reproduced with copyright permission from http://www.teradyne. com/flex/microFLEX.html)

There are several approaches to achieve testability and fault simulation in MEMS parts, but it becomes a common necessity to model both the MEMS and electronic sub-systems within the same simulation environment to ensure that faults are correctly introduced and analyzed [17]. During the development of the test solution, the test engineer might use a variety of tools and instruments to develop the test system solution but during the initial design phase it is necessary to evaluate features in the design (such as the test modes mentioned above) without an actual instrument and this is possible using a virtual instrument (such as IVI[12]) that allows the developer to run code without a physical instrument or component. IVI simulations are enabled through available APIs that standardize common measurement functions and simplify testing of measurement applications. One necessary component of these types of simulations is a system level representation of the MEMS device and associated electronics, usually available as a schematic model (described earlier).

Figure 7.8 depicts an example of a typical MEMS comb-drive capacitive accelerometer device represented in a schematic editor[13] and the simulator is able to analyze the response of the accelerometer to a variety of input stimuli. In this environment, fault analysis i.e. detecting malfunctions, would be performed in a closed loop configuration to capture the effects of non-idealities like process variations (see Fig. 7.4), noise, mode coupling, resolution limitations, etc. It is more efficient to perform these types of simulations and investigations during the design phase in a virtual environment then doing so after when a significant amount of time would have to be spent on developing measurement systems for every feature or mode that needs to be measured. The availability of models (e.g. Fig. 7.8) also makes it easier for the test solution (test code and hardware) to be developed and debugged early in the development.

A common field failure mode in MEMS is particulate contamination during fabrication; it is a major source of hard-to-detect failures in MEMS devices that can have a major impact on both manufacturing yield and field reliability. Particles of various sizes behave quite differently in terms of their mobility and effect, and to capture or measure the impact of such particles, it is possible to introduce a particle model into the system model by modifying the schematic shown in Fig. 7.8. This is accomplished by using a process simulator that inputs the sensor layout [18, 19], and creates a simulation (FEA) model to correlate the faulty behavior (i.e. performance with the particle) to the source of the contamination and outputs a netlisted model of the defective MEMS structure which is used at the test solution development level (IVI simulations). The FEA analysis is dependent on the size of the model and introducing such faults at the schematic level is the most efficient way to ensure complete test coverage prior to production. In [19], the impact of particles on a typical MEMS electrostatic comb-drive resonator has been demonstrated and a wide variety of mis-behaviors was observed to be caused by such particles.

Lastly, the ability to verify product behavior before silicon tape-out is of vital importance in achieving increased confidence in the design and improved yields and quality. The simulation of behavioral models (of the DUT) including the application hardware, tester resources (which maybe ideal models that are independent of the specific test platform) is necessary to understand and identify some of the complex correlations between the DUT and tester. Improved modeling capabilities that allow for introduction of various MEMS specific faults such as particles or stiction will enable much higher yields in final production.

---

[12]Interchangeable virtual instrument (IVI).

[13]Cadence – Spectre™ or Synopsys Saber™ are examples of schematic capture tools.

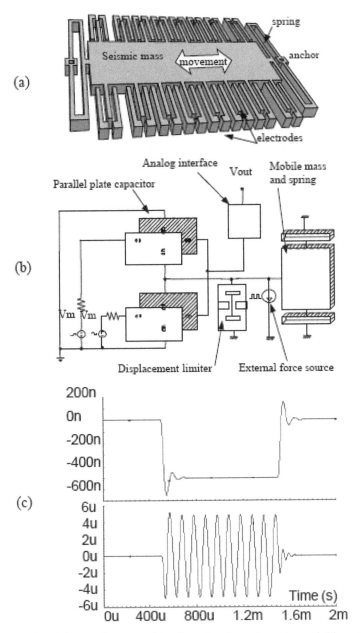

Fig. 7.8 (a) Capacitive accelerometer (b) schematic of accelerometer and (c) response of accelerometer to input pulse (reprinted with permission Copyright 2001 IEEE [17])

7.2.3 Process and Packaging Integration

To improve performance it becomes necessary to optimize the coupling between the micromachined element, the electronic control circuit and the package and so in this section we will examine integration of the process descriptions to maximize overall yield.

As a first step, it is possible to represent the micromachined element design as a single degree-of-freedom lumped model, which can be used by IC designers for design iterations of the ASIC or test engineers to virtually prototype the test system. These simple models are not capable of capturing enough of the fabrication or assembly level process detail making it challenging to investigate non-performance of the part at some specific tolerance limit but they do give designers enough insight to address the non-conforming parts of the design. For example, the temperature or voltage coefficients of the part can be simulated and targeted. Recent advances in the integration of sophisticated fully coupled reduced-order models libraries [13] with conventional IC design and manufacturing tools [20] now provide a single CAD environment for the design of both the MEMS element, and ASIC.

From a package or assembly standpoint, the design methodology is similar but it is not simple to include assembly process tolerances into the design process without having to couple several dissimilar design tools [21, 22]. Typical assembly process tolerances such as die alignment or tilt tolerance within a package or control of bond line thickness (due to dispensation of adhesive) are introduced because of the machines used to automate the assembly process such as a pick and place machines, wire bonders, epoxy dispensers etc.. The process tolerances in assembly are not as controlled to the same level as a clean-room process but at a coarse level the influence of these tolerances can be incorporated to investigate performance variations [23].

A *compact* model [21] of the package is a model that describes the mechanical deformation state of the package over a range of external loads such as temperature or other environmental influences. Such a model may be obtained by creating an FEA model of the package and simulating the thermo-mechanical deformation and stress behaviors over the range of environmental influences, package-induced effects, and history dependent effects (such as built-in residual stresses) experienced during the useful life of the part (see Fig. 7.9 for example).

The package compact model is also convertible to a schematic level model that could be "attached" to the MEMS schematic model in an analogous way to the attachment of the device within the physical body of the package i.e. a boundary condition. The main assumption here of course is that the MEMS element is small enough that the influence of the element on the package is negligible and under these conditions, the MEMS device is now subject to the same direct environmental and package induced stresses as experienced by the package.

Since the extracted package models are parametric i.e. they have process parameters that can be varied, the designer could also simulate the device anywhere in the external parameter space between the maximum and minimum parameter values that were used in the original package model extraction. As an example, the

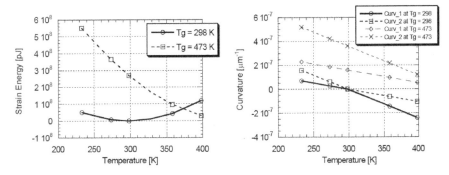

Fig. 7.9   Package Modeling – Estimation of the curvature of a die in a ceramic package at multiple temperatures (reprinted with permission Copyright 1999 NSTI [22])

frequency and capacitance of a MEMS accelerometer as a function of package stress is shown in Fig. 7.10. This methodology to simulate package effects on the performance of the MEMS device may also be extended to system level models [24] thus creating more sophisticated and robust designs that will have higher manufacturing yields.

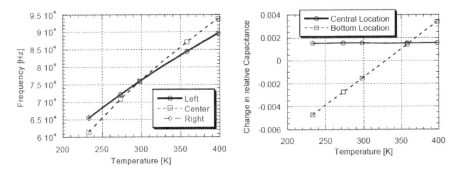

Fig. 7.10   Package modeling – device frequency and capacitance changes as a function of package temperature and die position (reprinted with permission Copyright 1999 NSTI [22])

As part of a yield improvement strategy, MEMS packaging needs to be developed and engineered at the same yield levels as the front-end fabrication. Often in a product development cycle, package design is handled by a separate team which is not engaged until late in the development cycle and this approach has been shown to be one of the primary cost drivers for MEMS [25]. Another possible strategy has been to use COTS[14] packages with standard form factors, sizes and pin-outs, which will keep costs down and minimize the need for additional tooling and process development.

---

[14]Commercial Of The Shelf (COTS)

7.2.4 Yield Modeling

Yield modeling in MEMS is still developing, with much of the research being conducted in the IC design space, and being applied to MEMS wherever possible [26]. To generate an accurate MEMS yield model it is essential to have accurate defect data (see failure modes and mechanisms in Chapter 3, and FMEA Section 5.2) for the technologies used in the product i.e. surface micromachining, wafer bonding, plastic over-molded packaging etc. In IC design, a common method of accurately counting defects is the *critical area method* (which counts the number of defects in critical areas of the chip layout) which requires defects to be classified in a particular way. Basically, the defect data required for a critical area based yield model is all of the defects present in the process that could cause a fault, irrespective of where they occur on the chip [27]. So even if the defect falls in a non-critical area (where it would not cause failure) it is still "counted" as a defect. This is because it is important to build up the defect statistics independent of the particular design.

As described earlier, MEMS defects are classified according to the failure mechanisms they cause and counting defects using an area method is less useful. However, all faults that affect the same parts of the design or influence the same types of parameters and thus cause the same kind of failure in a layer can be grouped together (also called fault classes [28]) or if they can be put into sub-groups counted together, the yield loss associated with the sub-group can be quantitatively described.

For MEMS yield modeling, the Monte Carlo method (introduced earlier) has been used [29] for the simulation of point-stiction defects in MEMS accelerometer devices. The yield of MEMS devices is estimated based on the comparison of simulated yields of a built-in self recoverable and non-recoverable MEMS accelerometer and demonstrated the possibility of increasing the effective yield with a self-repairable design. A similar approach was used [30] for the estimation of the yield of a MEMS RF switch due to process variations. The functional yield parameter of interest was the tip displacement (δ_{tip}) due to process tolerances in the various structural layers that make up the switch. The Monte Carlo method was used to set up functional relationships and set constraints and the resulting sensitivity analysis identified the particular layer or layers that influenced the tip displacement the highest. These parameters are then used to create the functional yield model shown below in Fig. 7.11.

Name	Dimension (μm)	Nominal Variation	Improved Process Variations [+/- %] (3σ)				
Sacrificial Layer	1.5	14	10	7.5	5	5	5
Sacrificial Layer2	0.5	12	12	12	5	5	5
Metal	0.5	7	7	7	7	5	5
Oxide Layer	2	15	10	7.5	5	5	5
Metal2	0.5	10	7	7	7	5	5
Tip Deflection	0.2	0.15	0.15	0.15	0.15	0.15	0.06
	Switch Yield %	83	92	95	98	98.5	99.5

Fig. 7.11 Functional yield modeling of an RF MEMS switch (reprinted with permission Copyright – 2003 ASME [30])

The utility of functional yield modeling prior to the actual fabrication of devices is only beginning to be felt in MEMS product development. This is because of several factors but the foremost has been the lack of suitable CAD tools that enable robust design through parameterized behavioral models within existing network simulators as well as a quantitative analysis of the failure mechanism within a given process. MEMS specific simulation tools [31] are beginning to address these needs but much work remains to be done. As MEMS becomes more widespread, this type of yield modeling will continue to gain popularity as it has undeniable benefits to cost and overall quality of MEMS products.

In the next section, we will begin to look at the other part of the quality equation, namely field reliability enhancement. The contributing factors to improving reliability are primarily fall into two product development categories i.e. design and manufacturing. We will look into the contributing factors more closely.

7.3 Reliability Enhancement

The previous section focused on improving yield which is one part of overall quality, and in this section we will see that there are additional areas within the design and manufacturing spheres that can influence the reliability of the product. In design, probably the most influential factor is reliability modeling which depends on both the CAD methodologies employed and through it on process and material characterization [31]. The manufacturing phase is broadly divided into 3 major consecutive activities, (a) process short loops, (b) prototype and demonstrator parts and (c) pre and volume production. In each of these sub-phases continuous improvement and refinement of the process must occur to further reduce variability and cost and improve reliability. By far, ensuring the stability and reproducibility of the process, collecting properly characterized properties, enforcing conservative design rules and the use of reliability stress methods (as described in Chapter 6) can lead to vast improvements of field reliability.

7.3.1 Process Stability and Reproducibility

Independent of the business model adopted a manufacturing unit that is involved with MEMS necessarily needs to ensure a stable and reproducible fabrication process to produce reliable product. Usually, a product that relies on the development of a new process flow (either during fabrication or assembly) may not achieve the highest level of reliability possible within that manufacturing process until after the part has been released to the market but subsequent products using the same manufacturing flow will no doubt benefit. Over time the process will mature to where it is possible to have extremely low field failure rates (a few ppm) and the challenge is to reduce the time to reach that level of field failures as soon as possible.

A process is said to be stable when all of the parameters that are used to measure the process have constant means, variances and distributions over time [32]. The path (Section 7.2.1) to achieving a high level of confidence in the manufacturing process is quite simple but time consuming – the more units processed the more stability and reproducibility can be achieved within the manufacturing process. For example, during fabrication process development, if there is need for either modifying a unit process step or changing the process flow; development begins with short loop experiments (DOE) to evaluate the effect of key process variables. Based on the results, more wafers are processes to verify or validate the result, and then even more wafers are processed to ensure reproducibility. This approach leads to verification of design rules and quantitative measurement of material properties, and ultimately a well qualified stable and reproducible process flow. However, this can take a very long time as it becomes necessary to process wafers, assembly parts, test, and analyze the resulting data. Lastly, a manufacturing process cannot be released to production (or evaluate process capability) until it has been proven to be stable, nor will it unless the foundry is capable of achieving a certain minimum level of fabrication yield.

7.3.1.1 Process Characterization

Fabrication process characterization test die (as shown in Fig. 7.12) are used in MEMS foundries for evaluating stability and reproducibility of unit process steps and the overall process flow [33]. The measurement of geometrical features on the test die including film thicknesses, gaps, line-widths, line spacing, and alignment features results in valuable data for the development of process design rules

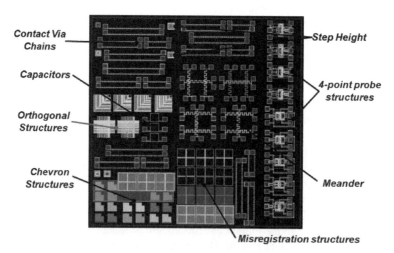

Fig. 7.12 Process characterization test die (reprinted with permission Copyright – 2001 Coventor)

whereas the electrical structures (capacitors, meander resistors) are useful for measuring electrical properties within various layers. Typical design rules that need to be defined and consistently measured include the minimum feature size, maximum etch hole separation, thickness of thin film materials, required overlap in layers, and others. For a given process, the list of design rules necessary must contain all information necessary to produce a working device.

Additionally, such die may also contain electrical parameter related devices such as fixed capacitors, 4-point probe structures and (meander) resistors, which are necessary for measurement of electrical permittivity (ε) or resistivity (ρ) in a particular layer or between layers. Once a process has been released to production, process related data from process characterization test chip becomes useful in creating process design kits [34, 35].

Lastly, the fabrication line can use such die as *process control monitors* (PCMs) during the production process as process controls based on some of these geometrical or electrical property measurements. PCM's yield specific information aloowing one to predict the performance with respect to the design specification. The PCM could be placed at several locations on every wafer and measured optically or electrically after each processing step or at the end of the fabrication line, based on the quality and quantity of control required, thus allowing for strict in-line process controls that maintains process quality.

MEMS fabricated on a qualified process with qualified test structures according to prescribed design rules will benefit greatly in terms of improved reliability. In addition to improved yield and reliability, the standardization of the process results in cost reduction.

7.3.1.2 Material Property Characterization

Since MEMS devices can be designed to transducer a variety of signals, their design requires accurately characterized material properties. The variety and diversity of MEMS makes it difficult to capture every single relevant material properties can be classified roughly into five groups – the three primary material property groups such as elastic, electrical and thermal properties and two reliability property groups such as loss mechanisms and failure mechanisms. Depending on the process steps involved, the specific material property of interest (i.e. modulus or CTE or damping coefficient) may be in fact a composite property or a combination of each such group. A summary of relevant geometrical parameters, material and reliability properties for MEMS are provided in Table 7.1.

Similar to the process characterization die in the previous section, material property characterization die (shown in Fig. 7.13) containing several different types of test structures [37] are placed at specific locations on a wafer. Many variables such as size of the wafer, test structure characteristics, and process characteristics, as well as availability of metrology equipment in the clean room will determine the placement of such characterization die and the types of structures within each die. Figure 7.13 shows a typical material property characterization die with several

Table 7.1 List of geometrical, material properties and reliability property groups (adapted from [36])

Geometry	Primary material properties	Reliability properties
Thickness	Elastic properties	Loss mechanisms
Undercut	Residual (intrinsic) stress	Material and gas
Line width spacing	Stress gradients	Damping
Sidewall profile	Anisotropic modulii	Contact between
Step coverage	Non-linear elasticity or	Surfaces
Radii of curvature	plasticity	Cavity pressure and
Voids	Piezoelectric coefficients	hermiticity
Minimum feature sizes	Electrical properties	Failure mechanisms
Misalignments	Dielectric permittivity	Yield and fatigue
Surface roughness	Electrical conductivity	strength
	Frequency dependent	Delamination / bond
	properties	strength
	Thermal properties	Creep
	Anisotropic CTE	
	Specific heats,	
	Thermal conductivity	

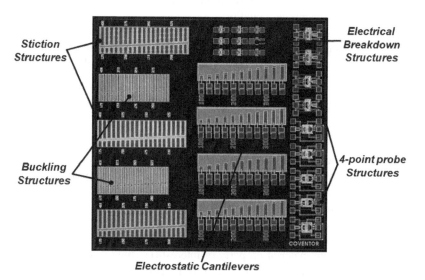

Fig. 7.13 Material property characterization die (reprinted with permission Copyright 2001 Coventor)

different kinds of test structures for measuring stiction between layers and buckling, as well as cantilever structures for measuring electrostatic pull-in.

The specific objective of having such die is to measure intrinsic material properties such as the Young's modulus, fracture strength, electrical conductivity, dielectric permittivity, and TCE or extrinsic properties such as stress and stress gradient, but there are quite a few additional properties of interest depending on the specific product application in mind [38]. Since there are many different types of test structures, discussion of each type is not possible in this chapter and so instead, we will discuss only a few of these test structures and the measurement techniques used in detail.

7.3.1.3 Test Structures and PCMs

To enhance reliability, test structures to characterize material properties must be designed within the process of interest [36]. It is these test structures that will give the product team a quantifiable metric that the design will perform as expected in the field. This section describes some of the basic test structures used and how they measure a particular property.

Similar to dimensional metrology, the measurement of thin-film electrical properties such as sheet resistance, temperature dependent resistivity, and dielectric constant, use well-established techniques; MEMS fabrication units have largely adopted the same techniques and reader is referred to other sources [39]. These techniques include the use of 4-point probes, meanders, contact-via chains and capacitors [40]. In this section we will review some of the techniques and methods currently employed for measuring both thermal and mechanical material properties of MEMS materials.

There are many properties (Table 7.1) and some of them may be more important for a particular product relative to another but in this sub-section we will only look at a few properties that are most commonly encountered.

Elastic Modulus and Strength

The measurement of elastic modulus at the wafer level may be accomplished in a variety of ways although most techniques focus on measuring a single elastic modulus (the Young's modulus) and are limited in terms of measuring anisotropic modulii unless specific modifications are implemented. In [41], a detailed summary of anisotropic values of Young's modulus, Poisson's ratio, and other elasticity quantities is provided, which is useful for design and analysis of silicon based structures and should enable designers to use more accurate descriptions of the modulus in design calculations. The benefits of using the full anisotropic description of modulus especially in complex designs can be quite substantial. To date, there are several ex-situ techniques (outside the fab environment) that are based on miniaturized versions of established test methods (e.g. tensile test, micro-hardness test etc.). A summary of demonstrated measurement techniques for Young's modulus is provided in Table 7.2.

A comparison of the various methods reveals the diversity of demonstrated methods for measuring the Young's modulus and strength of thin film polysilicon, nitride, and metal, and that more measurements have been reported on Young's modulus (E) of silicon and polysilicon than other materials. Although a sub-set of these techniques have the capability to measure complimentary properties such as the

Table 7.2 Methods for measuring the modulus of elasticity and strength (adapted from [31])

Test type	Properties measured	Requirements/issues	Notes/description
MicroTensile Test	$\sigma-\varepsilon$ curve E, ν, σ_y and σ_f [32–35]	Specimen preparation Delicate handling Alignment Dimensions of sample Calibrated load Strain measurement	Interferometric strain/displacement gauge [36] AFM based [32] SEM based [33]
Microstructure bending test	E, σ_y and σ_f [37–40]	Specimen preparation Alignment Sample dimensions Calibrated load Deflection measurement	Mechanical loading of cantilevers and fixed-fixed beams
Active Test structures [41]	E, σ_y and σ_f [42, 43]	Sample dimensions For resonant structures mass density also needs to be known (or one could add a large known mass)	Resonant beams [44] Resonant comb device [45]
Bulge Test	$E/(1-\nu)$ [46] ν [47] Measurement of σ_y and σ_f demonstrated [48, 49]	Dimensions of membrane (i.e. thickness, in-plane) Accurate pressure measurement and control	Optical white light interferometry [46] Michelson interferometer [47]
Ultrasonic measurement	K or $E/(1-2\nu)$	Mass density and Poissons ratio of material must be known	Optical methods [50]
Nanoindentation on blanket films	H, E, and σ_y [51, 52]	For Anisotropic materials modulus obtained is some average of the moduli in different crystallographic directions. For isotropic materials, formulas are available for the relationship between this derived modulus, and the Young's modulus and Poissons ratio. For Textured Films the determined convoluted modulus has a limited association with either Young's modulus or any other modulus.	Load-Displacement curve automatically generated from instrument

yield stress (σ_y) and failure strength (σ_f), these properties have not been as widely reported.

The basic method involved in these modulus measurement techniques is the generation of a stress-strain curve from which the modulus is extracted. Numerous articles [46, 44, 50, 42, 58] have demonstrated the accuracy and validity of *microtensile* and *microbending* as direct methods for measurement of modulus or the alternate *Bulge* test [55, 56, 61] technique which is an inverse method that requires curve fitting of the load-displacement curve to the theoretical behavior of the membrane. Ultrasonic methods with ultra-short laser pulses [59] offer an option as a non-contact method while nano-indentation is a destructive contact method but still very useful when correctly used and interpreted [61] but the destructive nature of the test and the requirement of a specialized instrument make its adoption as an inline tool quite challenging. Also, *active* or resonant test structures have fairly wide acceptance in terms of simplicity, minimal specimen handling and electrical output, and is approved as a standard for integrated MEMS processes [38].

Finally, one should keep in mind that most modulus measurement techniques while capable of accurately measuring the modulus or strength are not really *inline* techniques, and are primarily useful as *ex-situ* techniques to corroborate in-line measurements.

Residual Stress/Strain

Residual stress is the most influential single property in thin films and there are several established techniques to measure residual stress:

1. Analytical Methods: The predominant technique is *Raman spectroscopy* which again is an *ex-situ* technique that allows a stress measurement within the first few monolayers beneath the surface with minimal specimen preparation. There have been demonstrations of other methods such as FTIR & X-ray diffraction [62] for MEMS materials and some of these techniques have been discussed in more detail in Section 5.5.

2. Wafer curvature [25] is routinely used evaluate the residual stress in blanket deposited films on wafers, and only requires the biaxial modulus of the substrate, and thicknesses of the film (t_f) and substrate (t_s) respectively. The *Stoney* formula (7.1) is used to estimate the biaxial residual stress:

$$\sigma_R = \frac{1}{6}\left[\frac{E}{(1-\upsilon)}\right]_s \frac{t_s^2}{t_f} \kappa \qquad (7.1)$$

Where subscript s indicates the substrate and f the film, and κ is the measured curvature.

3. Passive Microstructures: There are a variety of methods based on passive microstructures [63], such as fixed-fixed beams used for determining compressive stress, and microrings [64] for tensile stress. Most methods need a model to translate the direct measurand(s) into the parameter(s) of interest and one usually implements geometry such that the expected measurand range is within the measurement range. The microrings for instance, yield a critical *length/diameter* for which the test structure buckles, and thereby yield an approximate measure for residual strain.

Stress or Strain Gradient

Passive cantilever structures are the most convenient method for determining strain gradient given by the curvature:

$$\frac{\Delta \varepsilon}{t} = 2 \frac{d_{\text{free\_edge}}}{L^2} \tag{7.2}$$

If the deflection the cantilever is exclusively due to curvature (see Fig. 7.14), and constant along the length then any free end or convex corner of the structure can be used for the measurement. ASTM standards[15] are now available for measurement of in-plane length, strain and strain gradient of passive test structures assuming that the modulus of the material is known.

A modified version of the cantilever structure with electrostatic loading of an array of cantilever beams of various lengths provides an automated electric read-out of curvature effects [38].

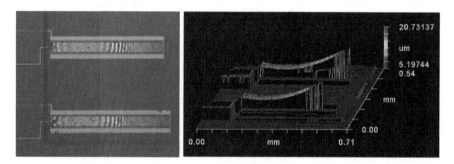

Fig. 7.14 Cantilever test structure for stress/strain gradient measurements (*left* – fringe structure observed due to curvature, *right* – 3D curvature measurement)

---

[15]ASTM E2244-02 – Standard test method for in-plane length measurements of thin, reflecting films using an optical interferometer.
ASTM E2245-02 – Standard test method for residual strain measurements of thin, reflecting films using an optical interferometer.
ASTM E2246-02 – Standard test method for strain gradient measurements of thin, reflecting films using and optical interferometer.

Creep and Fatigue Strength

Measurement of fatigue strength and temperature dependent creep for MEMS materials has been generally limited to metal MEMS with tensile LIGA-nickel specimens [65] one of the most studied material. Additionally, nanoindentation and the bulge test [55] have also been reported to investigate the creep properties of blanket deposited films of metals.

Fatigue strength is determined from dedicated resonant structures by measuring the number of cycles to failure [66]. As discussed in Chapter 4, fatigue is caused by crack growth during repetitive loading and for ductile materials the plastic zone around a crack tip is a few orders larger than the radius of the crack tip. In MEMS, these plastic zones could easily be larger than characteristic dimensions of the structure, and hence fracture either occurs during the first few loading cycles, or does not occur at all. A good review of fatigue testing at the micro-scale is provided by [67].

Quality Factor

The quality factor of a mechanical structure is dependent on the ratio of total damping to the critical damping and although it is not a fundamental material property, it is a loss property (as seen in Table 7.1) that quantifies dissipation in the material at existing environmental conditions such as the ambient pressure and temperature. Electrically addressed microstructures (such as cantilever beams, comb resonators, or fixed-fixed beams) yield information on the cavity pressure through response time and Q-factor measurement [68]. The main limitation with these approaches is that they all require accurate measurements of the geometry of the resonator as well as the modulus and density in order to calculate the quality factor.

Charging Effects

In electrostatically actuated MEMS test structures, charging of some kind will always occur but the severity of the charging really depends on the materials used. Charging as a mechanism depends on the presence of trapping sites, and for a conductive path to these trapping sites. Trapping sites may occur at the interface between two different materials (including between a solid and gas), or at grain boundaries or dislocations or other lattice defects. A path with poor conductivity will lead to long charging times, but will also interfere with charge dissipation which may lead to a charge build up. The amount of charging is determined from the pull-in voltage and $C(V)$ curve and in resonant microstructures charging is determined from its impact on the resonant frequency [69].

Stiction

Although "stiction" force is not an intrinsic material property it is a cumulative measure of the effects of processing on the surface adhesion energy (see Chapter 3). Among the first in-line approaches using test structures to measure stiction was based on the detachment length of an array of cantilevers of varying length [70], which can be correlated to the adhesion energy of the surfaces in contact. An ex-situ technique for measuring the adhesion energy using an AFM has been reported [71].

In summary, material properties are critically important for understanding the interaction of the process and the design and the overall impact on reliability. The material property information that has been previously correlated with simulation models of the test structures should be made available in process design kits and utilized by designers to predict performance of the device over a wide variety of operating conditions.

7.3.2 Product Qualification

Product qualification for MEMS products is highly dependent on the end application and the packaging that houses the MEMS die and electronics. In Section 6.4, examples on the qualification protocols for a typical airbag accelerometer and Tire Pressure Monitoring (TPMS) sensor show that the process of qualifying a MEMS based product presents a major challenge to make sure that the qualifying stresses appropriately target the potential failure modes identified in the FMEA. A significant part of enhancing reliability is the ability to incorporate the learning from the look-ahead qualification or stress testing into the design cycle so that parts are designed with reliability in mind. As we have seen in Chapter 2, acceleration testing is invaluable in identifying potential failures in a reasonable amount of time. Before we discuss design for reliability for a particular product we should understand a little more about acceleration factors (Section 2.2) and how the reliability stress plan is created.

7.3.2.1 Acceleration Factors

In reliability statistics (see Section 2.1) a simple failure rate (λ) calculation based on a single life test may be described using Equation 7.3:

$$\lambda = \frac{1}{a_F \times H} \qquad (7.3)$$

where λ is the failure rate, a_F is the acceleration factor,[16] and H is the total device hours.[17] As seen in Fig. 2.2, the failure rate follows a typical bath-tub curve *infant mortality* in early life, a stable rate during useful life, and an increased rate during *wear-out*. The a_F allows the stresses used in the reliability testing to accelerate the time to failure of a particular failure during field use.

---

[16] Acceleration Factor (a_F) is a constant derived from experimental data that relates the times to failure at two different stresses – see Section 2.2.

[17] Total device hours (H) is the summation of the number of units in operation multiplied by the total time of operation.

As described in Chapters 3 and 4, a MEMS part may have several potential failure mechanisms and so a comprehensive failure rate (λ) is required but calculating such a compound failure rate can be challenging since failure mechanisms are thermally activated at different rates. Equation (7.4) (similar to (2.23)) below accounts for these conditions and includes a statistical factor to obtain the confidence level for the resulting failure rate.

$$\lambda = \sum_{i=1}^{N} \left(\frac{x_i}{\left(\sum_{j}^{K} a_{Fij} \times H_j \right)} \right) \times \frac{M \times 10^9}{\sum_{1}^{N} x_i} \tag{7.4}$$

Where N is the number of distinct failure mechanisms, and K is the number of life tests being combined, and x_i is the number of failures for a given mechanism. In the failure rate calculation, acceleration factors (a_{Fij}) which are calculated using the Arrhenius equation, are used to de-rate the failure rate from the thermally accelerated life test conditions to a failure rate indicative of actual use temperature. The activation energies for certain MEMS failure modes is shown in Table 7.3 and these may be used in particular acceleration models (Tables 2.2 and 2.3) but the availability of such activation energies estimates requires comprehensive experiments with production ready devices.

As an example, the feasibility of using temperature and humidity to age vapor deposited SAM-coated electrostatic-actuated MEMS devices with contacting surfaces in torsional ratcheting actuator was demonstrated by [72]. Failures were seen

Table 7.3   Activation energies of MEMS failure modes

Failure mechanism	Activation energy	Screening and test methodology	Control methodology
Bulk silicon defects	0.3–0.5 eV	HTOL	SPC on thermal processes
Assembly defects	0.5–0.7 eV	Temperature cycling, temperature and mechanical shock	SPC on assembly process
Silicon fracture (mechanical stress)	0.8–1.0 eV	Vibration and temperature	
Mask defects/photoresist defects	0.7 eV	HTOL, defect density monitors	SPC control of photoresist/etch process
Contamination	1.0 eV	HTOL	Clean fab and assembly processes

to be dependent on both temperature and humidity and measured an equivalent surface damage at shorter time intervals and higher humidity compared to longer times
at lower humidity.

7.3.2.2 MEMS Qualification Testing

In the previous section we saw how Accelerated Life Testing (ALT) plays a critical
role in causing failures in time-scales that are favorable for product development.
The failure activation energies (Table 7.3) and mechanisms shown (Table 2.2) show
that it is possible to design the qualification stress tests to trigger or probe a particular failure mode. As part of the reliability development and life improvement of the
MEMS part, it is important to begin life testing as early as possible. Ideally, the first
generation of devices and test structures would be subjected to *look-ahead* testing
using experiential knowledge (captured through the FMEA) of the failure mechanisms to probe the reliability and performance of the part. As described in [73],
the necessary steps to execute ALT include collecting historical data (on materials, processes, field returns), accelerated life testing, failure analysis, statistical data
processing, corrective actions and predictive methods.

Reliability or stress testing of MEMS prototypes is also important because it
can (and does in many cases) uncover potential failure mechanisms not considered
in the FMEA process, and in some cases this leads to further process or design
modifications which can have important schedule or design consequences. On subsequent iterations the results of stress testing will improve because of design or
fabrication refinements introduced for mitigation of a particular predominant failure mechanism until finally as the device transfers into full–scale production there
is a marked improvement in reliability (Fig. 7.15 below) brought about by all

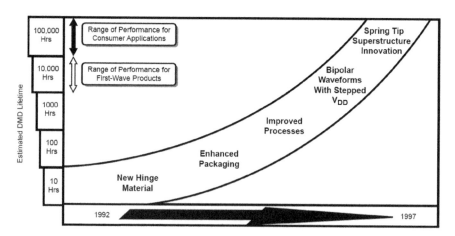

Fig. 7.15 Reliability development and life improvement (reprinted with permission Copyright
IEEE [1])

design innovations, production processing, and testing improvements. These reliability fluctuations from design to production can be minimized by forcing reliability improvements to coincide with device design and production; which will ultimately lead to a more reliable device with a faster time to market. In the case of the DMD® mirror,[18] the improvement in reliability to operating regimes appropriate for consumer display applications was achieved by design innovations (i.e. design for reliability – DfR) which will be discussed in Section 7.5.

Product qualification encompasses a set of simulations, predictive methods, and measurements to establish the mechanical, electrical, thermal, and reliability characteristics of a particular device. In Section 6.4, a detailed description on quality standards including MIL, Telcordia, and automotive standards was introduced, through specific examples on quality testing protocols for MEMS accelerometers (Fig. 6.42) and other MEMS products. During the development of any MEMS product, the DFMEA and the reliability requirements guide the development of the reliability or quality plan and the number of samples, stresses and times are usually decided by a process illustrated in the example discussed next.

Example – Assume that the reliability requirement for a new MEMS part is 99% survival for 10 years under normal operating conditions at a 95% confidence level. The failure rate is calculated from Equation 7.5:

$$\lambda = \frac{\chi^2_{(1-\alpha,2r+2)}}{2t} = \frac{\chi^2_{(1-\alpha,2r+2)}}{2Ta_F} \tag{7.5}$$

where χ^2 is the chi-square function, α is the confidence level, r is the number of failures, t is the device hours at 25°C, and T device hours at the acceleration temperature 125°C (say).

If we assume an activation energy = 0.5 eV (a bulk silicon or assembly defect for instance), the acceleration factor at 125°C works out to be:

$$a_F = \exp((0.5eV/8.62 \times 10^{-5}eV/K)(1/298K - 1/398K)) = 133.03$$

The total number of device hours "T" required to achieve 99% survival rate for ten years (or 87600 hrs), under normal operating conditions at a 95% confidence level, is derived as follows:

$$\lambda = \frac{(0.01)}{87600} = 1.14 \times 10^{-7} \text{failures/h}$$

---

[18]See Section 2.5.1.

So for 0 failures (for example) at the acceleration temperature[19] the total test device hours is:

$$T = \frac{5.99}{(2)\,(133.03)\,\left(1.14 \times 10^{-7}\right)} = 197,488\ \text{h}$$

If we use a sample size of 200 samples, the test duration is then 197488hs/200 samples = 1002 hrs/sample. In other words, we have to test 200 devices for a 1000 hrs each with no failures.

In this way, the reliability plan including stress levels and test times are calculated and the sample test plans shown in Chapter 6 are developed. In the next section, we will consider the topic of design for reliability (DfR) which is another component of the overall effort to enhance reliability. As has been discussed in the case of the DMD® mirror (and shown in Fig. 7.15), design changes are a significant contributing factor to improved reliability and we will try to understand how incorporating reliability into the design process can be realized in MEMS.

7.4 Design for Reliability (DFR)

Through references cited in earlier chapters (such as [1]), there is considerable evidence that MEMS can be made with a high degree of reliability and examples discussed earlier (Section 2.6) such as Texas Instruments DLP®, ADI accelerometers, RF switches, add credence to this observation. Additionally, over the past two decades, there has been a vast amount of research into the fundamental failure mechanisms of MEMS materials (Chapter 3). However, it can still be a relatively major challenge to design MEMS device with a known time to failure or clear prediction of *safe* operating regimes. Among the main reasons for this challenge is the availability of an integrated system level predictive design methodology to create and apply reliability knowledge to MEMS designs.

Design for Reliability (DFR) is a discipline that refers to the process of designing reliability into a product [31]. This requires that the product's reliability requirements be clearly defined before it can be "designed in" to the component or system. In general, DfR comprises of five essential but different integrated areas most of which have been covered in this book.

- Methodology: In the context of DfR, the methodology adopted to design reliability into the part must occur at two different levels. At a basic level (as identified above) a product's key reliability specifications must be clearly understood. Whether a part is to survive for 2 years or 10, without a clear target it is not possible to design in reliability. At the other level, methodology also includes

---

[19]The factor 5.99 is the 95th percentile of the χ^2 with 2 degrees of freedom which corresponds to 0 failures.

the organization's approach to reliability enhancement. For example, if there is a known susceptibility in the fabrication process to delamination between two adjacent layers, or an observed performance shift during accelerated testing, there must be motivation and a methodology to drive to root cause analysis so that appropriate steps can be taken to improve the technology for future products. Lastly, at the part development level, there should exist a clear methodology to go from reliability study [74] of initial prototypes to final release of the product to the market (see case studies in Chapter 2).

- Design Tools: In order to be able to predict failure modes during life it becomes necessary to be able accurately simulate the physics underlying the root cause of failure. In earlier chapters we have seen that MEMS designers use a variety of commercially available tools from system level analysis such as network level simulators, and Matlab$^{\circledR}$/Simulink$^{\circledR}$ level tools to field analysis tools including FEA, BEM and others. These tools are used quite effectively in MEMS design to model a variety of failure modes such as fracture or delamination during high g shock, contact wear, and stiction.

- Metrology: The need for metrology i.e. accurate property measurement, is necessary both at the level of well defined experiments or test structures (as we recall from Sections 7.3.1.1 and 7.3.1.2), as well as at the level of failure analysis methods (discussed in Section 5.5). The ability to accurately measure process variations and tolerances enables data analysis to detect small shifts in performance that might otherwise be undetected. A simple example is the residual stress in a film, which can change over time and temperature. The ability to measure this stress as accurately as possible gives a much more accurate prediction of expected performance over time or other operating conditions.

- Materials Science: At some level, improving reliability is primarily concerned with the material science of fairly unique materials involved in MEMS devices. From elastic brittle materials like polysilicon, to epoxy based die-attach in conventional plastic packages, to intermetallics involved in solder reliability, an understanding of the root cause involved with a particular failure mode necessarily depends on a good understanding of the material science involved. A further consideration is that the same material produced in different machines may well end up with slightly different properties.

- Verification: A major component of designing for reliability is the ability to incorporate learning from earlier stages in development. In order to do so, reliability must be statistically verified (or the risk involved must be accepted). This not only means that the reliability test strategy including accelerated test protocols (Sections 2.2 and 7.3.2.2) be implemented and executed, but that prototype demonstration and any compliance testing be carried out prior to release.

In terms of implementation of these components, reliability based analysis begins with the development of a suitable MEMS model (as shown in Fig. 7.4) with specific failure modes incorporated in the form of components within the model. The failure mode components monitor the failure point by implementing a known failure model that is activated when certain conditions are met or when the component exceeds a set of specified limits. These failure models would contain the specific reliability science that is specific to the production process used to manufacture the device.

As an example of a particular failure mode implemented in design we can look at the fracture failure of a MEMS device made with a SiC thin-film. The fracture strength of MEMS devices is known to be affected by the surface defects and surface roughness resulting from the manufacturing process. Such variability can directly impact the failure modes and, in turn, the reliability of the device. A probabilistic model [75] describing the fracture strength behavior of brittle materials (Fig. 7.16) could be incorporated into the system level MEMS design that could then predict reliability performance of the part. Reliability based optimization methods have been developed for MEMS devices [76] and while they are not yet ready to replace deterministic optimization methods they provide a useful tool to design under uncertainties.

Fig. 7.16 Weibull plots (P_f, probability of failure) for ultra-nano crystalline diamond (reprinted with permission Copyright 2003 JMPS [77])

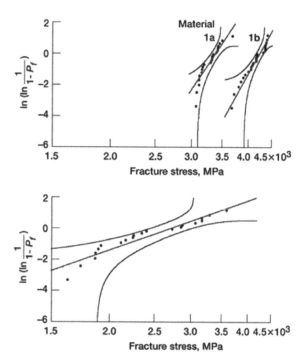

In essence, the design technique employed i.e. based on the physics of failure relies on a detailed understanding of the physical process of stress, material behavior, temperature and failure for that specific process. There are other design techniques that are common in DfR such as redundancy or derating [78] and these could be modeled in the same CAD framework. Redundancy is the design-in of a redundant or backup system or component and while it is possible to significantly improve the reliability of a system using this technique, the approach adds cost. Derating on the other hand is similar to adding a factor-of-safety by designing a component or system whose stress tolerance significantly exceeds the requirement.

7.5 Summary

MEMS reliability is a challenging topic to describe without a specific application case or example to work through. The myriad number of applications, from pressure sensors to accelerometers to mirrors and RF switches, each with their own challenges and goals for improving reliability, are definitely too numerous to cover in a single book. However, the main focus of this chapter is on methodology for improvement of reliability of MEMS designs and attempts to delineate the concepts of yield from reliability as they relate to overall quality of the product introduced to the market. For product developers, the main message is to not to rely on the accuracy of fabrication and to build designs that are robust to batch and foundry variations. Employing test structures to quantitatively understand the interaction between the process and the design and its impact on reliability is critical.

References

1. Douglass, M.R. 1998, Lifetime estimates and unique failure mechanisms of the Digital Micromirror Device (DMD), Reliability Physics Symposium Proceedings, 1998, IEEE pg 9–16.
2. Schropfer, G., et al. 2004. Designing manufacturable MEMS in CMOS compatible processes – methodology and case studies. Strasbourg, France: SPIE, SPIE Photonics Europe – MEMS, MOEMS and Micromachining.
3. da Silva, M.G., et al. 2002. MEMS Design for Manufacturability. Boston: Sensors Expo.
4. Tennant, G. 2001. Six Sigma: SPC and TQM in Manufacturing and Services. Aldershot: Gower Publishing, p. 25. 0566083744.
5. Anderson, D.M. 1990. Design for Manufacturability. Lafayette, CA: CIM Press.
6. Bralla, J.G. 1998. Design for Excellence. London UK: McGraw-Hill Book Co.
7. Romanowicz, B., et al. 1999. A methodology and associated CAD tools for Support of concurrent design of MEMS. Proceedings of the IFIP TC10/WG10.5 Tenth International Conference on Very Large Scale Integration: Systems on a Chip, Vol. 162, pp. 636–648. ISBN:0-7923-7731-1.
8. Lorenz, G., Morris, A., and Lakkis, I. 2001. A Top-Down design flow for MOEMS. Cannes, France: MEMS/MOEMS, pp. 126–137.
9. Romanowicz, B., et al. 1997. VHDL-1076.1 modeling examples for microsystem simulation. Toledo, Spain: 2nd Workshop on libraries, component modeling and quality assurance, accompanying hardware description language symposium.
10. Lorenz, G., and Neul, R. 1988. Network-type modeling of micromachined sensor systems. Proceedings of 1998 International Conference on Modeling and Simulation of Microsystems, Semiconductors, Sensors & Actuators, Santa Clara, CA, pp. 233–238.
11. Teegarden, D., Lorenz, G., and Neul, R. 1998. How to model and simulate microgyroscope systems. IEEE Spectrum 66–75.
12. Neul, R., Becker, U., Lorenz, G., Schwarz, P., Haase, J., and Wunsche, S. 1998. A modelling approach to include mechanical components into the system simulation. A Modelling Appro Proceedings of the DATE, pp. 510–517.
13. Coventor. 2009. Architect Reference. Cary, NC: Coventor, pp. T4–94.
14. Madou, M.J. 1997. Fundamentals of Microfabrication. Boca Raton, FL: CRC Press (Graduate level introduction to microfabrication. Fifth printing, Fall 2000), 2000.
15. Felton, L.E., et al. 2004. Chip scale packaging of a MEMS accelerometer. Electronic
16. Lubaszewski, M., Cota, E.F., and Courtois, B. 1998. Microsystems testing: an approach and open problems. Proceedings of the DATE, pp. 524–529.

17. Charlot, B., et al. 2001. Electrically Induced stimuli for MEMS self-test. VLSI Test Symposium. ISSN 1292-8062.

18. Kolpekar, A., Kellen, C., and Blanton, R.D. 1998. MEMS fault model generation using CARAMEL. IEEE. Proceedings of the IEEE International Test Conference, pp. 557–566.

19. Deb, N., and Blanton, R.D. 2001. High-Level Fault Modeling in Surface-Micromachined MEMS. No. 1/2, Dordrecht: Kluwer, Oct/Nov 2001. Journal on Analog Integrated Circuits and Signal Processing 29:151–158.

20. Schropfer, G., et al. 2003. Co-design of MEMS and electronics. 10th International Conference on Mixed Design of Integrated Circuits and Systems, Lodz, Poland.

21. McNeil, A.C. 1998. A parametric method for linking MEMS package and device models. Proceedings of the Solid-State Sensors and Actuators Workshop, Hilton Head Island, SC, pp. 166–169.

22. Bart, S.F., et al. 1999. Coupled package-device modeling for microelectromechanical systems. Technical Proceedings of the Second International Conference on Modeling and Simulation of Microsystems (MSM99), Vol. 40, pp. 232–236. ISBN: 0-9666135-4-6.

23. Hsu, T.R. 2006. Reliability in MEMS packaging. San Jose, CA: 44th International Reliability Physics Symposium.

24. Zhang, X., Park, S., and Judy, M.W. 2007. Accurate assessment of packaging stress effects on mems sensors by measurement and sensor–package interaction simulations. Journal of Microelectromechanical Systems 16:639–649, 1057-7157.

25. Senturia, S.D. 2001. Microsystem Design. Dordrecht: Kluwer.

26. Vudathu, S.P., Boning, D., Laur, R., and Phoenix, A.Z. 2007. A Critical Enhancement in the Yield Analysis of Microsystems. 45th IEEE International Reliability Physics Symposium (IRPS).

27. ITRS. 2009. International Technology Roadmap for Semiconductors: Yield Enhancement. ITRS.

28. Castillejo, A., et al. 1998. Failure Mechanisms and Fault Classes for CMOS-Compatible Microelectromechanical Systems. Proceedings of the International Test Conference, pp. 541–550.

29. Xiong, X., Wu, Y.-L., and Jone, W.-B. 2007. MEMS yield simulation with monte carlo method. In: T. Sobh, et al. (eds.), Innovative Algorithms and Techniques in Automation, Industrial Electronics and Telecommunications. Dordrecht: Springer Netherlands, pp. 501–504.

30. ASME. (2003). Course 469: MEMS Reliability Short Course. ASME.

31. da Silva, M.G., et al. Designing MEMS for Reliability.

32. Rogers, T., et al. 2005. Improvements in MEMS gyroscope production as a result of using in situ, aligned, current-limited anodic bonding. Science Direct Elsevier, pp. 123–124. Sensors and Actuators A, pp. 106–110.

33. Hembree, B. 2010. Production Process Characterization. NIST/Sematech Engineering Statistics Handbook. [Online] NIST/Sematech, 2010. http://www.itl.nist.gov/div898/handbook/ppc/ppc.htm

34. INTEGRRAM Metal-Nitride Prototyping Kit – Design Handbook. 2003. Metal-Nitride Surface Micromachining. QinetiQ Ltd. & Coventor Sarl.

35. McNie, M.E., et al. 2000. Software-Based Design Kits for Improved Microsystems Design in a Verified Simulation Environment. Copenhagen (Denmark): Eurosensors XIV, pp. 715–718.

36. Bouwstra, S., da Silva, M.G., and Schröpfer, G. 2003. Towards Standardization of MEMS Materials Characterization. Coventor. White Paper.

37. da Silva, M.G., and Bouwstra, S. 2007. Critical Comparison of Metrology Techniques for MEMS. San Jose, CA: Internation Society for Optical Engineering. Proceedings – SPIE The International Society for Optical Engineering, p. 646.

38. Osterberg, P.M., and Senturia, S.D. 1997. M-TEST: a test chip for MEMS material property measurements using electrostatically actuated test structures. JMEMS 6:107–117.

39. Schroder, D.K. 2006. Semiconductor Material and Device Characterization. Wiley. 0-471-73906-5.
40. Pineda de Gyvez, J., and Pradhan, D. 1999. Integrated Circuit Manufacturability – The Art of Process and Design Integration. Standard Publishers Distributors IEEE Press.
41. Hopcroft, M.A., Nix, W.D., and Kenny, T.W. 2010. What is the young's modulus of silicon? Journal of Microelectromechanical Systems 19.
42. Chasiotis, I., and Knauss, W.G. 2002. A new microtensile Tester for the study of MEMS materials with the aid of atomic force microscope. Experimental Mechanics 42:51–57.
43. Haque, M.A., and Saif, M.T.A. 2002. In-situ tensile testing of nano-scale specimens in SEM and TEM. Experimental Mechanics 42:123–128.
44. Sharpe, W.N. 2003. Murray lecture: tensile testing at the micrometer scale: opportunities in experimental mechanics. Experimental Mechanics 43:228.
45. Oh, C.S., et al. 2005. Comparison of the Young's modulus of polysilicon film by tensile testing and nanoindentation. Sensors and Actuators A:151–158.
46. Sharpe, W.N. 2002. Mechanical properties of MEMS materials. The MEMS Handbook 3:1–33.
47. Ericson, F., and Schweitz, J A. 1990. Micromechanical fracture strength of silicon. Journal of Applied Physics 68:5840–5844.
48. Johansson, S., Ericson, F., and Schweitz, J.A. 1989. Influence of surface coatings on elasticity, residual stresses, and fracture properties of silicon microelements. Journal of Applied Physics 65:122–128.
49. Weihs, T.P. 1988. Mechanical deflection of cantilever microbeams: a new technique for testing the mechanical properties of thin films. Journal of Material Research 3:1. Weihs, T.P., et al. "Mechanical Deflection of Cantilever Microbeams A:N931–942.
50. Zhou, Z.M., et al. 2005. The evaluation of young's modulus and residual stress of copper films by microbridge testing. Sensors and Actuators A127:392–397.
51. Gupta, R.K. 2000. Electronically probed measurements of MEMS geometries. Journal of Microelectromechanical Systems 9:380.
52. Allen, M.G., et al. 1987. Microfabricated structures for the in-situ measurement of residual stress, young's modulus, and ultimate strain of thin films. Applied Physics Letters 51:241–243.
53. Petersen, K.E. 1978. Dynamic micromechanics on silicon: techniques and devices. IEEE Transactions on Electron Devices ED-25:1241–1250.
54. Kim, Y.J., and Allen, M.G. 1999. In-situ measurement of mechanical properties of polyimide films using micromachined resonant string structures. IEEE Transactions on Components and Packaging Technologies 22:282–290.
55. Gally, B., et al. 2000. Wafer scale testing of MEMS structural films. MRS Symposium Proceedings 594.
56. Vlassak, J.J., and Nix, W.D. 1992. A new bulge test technique for the determination of young's modulus and poisson's ratio of thin films. Journal of Materials Research 7:3242–3249.
57. Small, M.K., and Nix, W.D. 1992. Analysis of the accuracy of the bulge test in determining the mechanical properties of thin films. Journal of Materials Research:1553–1563.
58. Edwards, R.L., Coles, G., and Sharpe, W.N. 2004. Comparison of tensile and bulge tests for thin film silicon-nitride. Experimental Mechanics 44:49–54.
59. Profunser, D.M., Vollman, J., and Dual, J. 2004. Determination of the material properties of microstructures using laser based ultrasound. Ultrasonics 42:641–646.
60. Oliver, W.C., and Pharr, G.M. 1992. An improved technique for determining hardness and elastic modulus using load and displacement sensing indentation experiments. Journal of Materials Research 7:1564–1583.
61. Saha, R., and Nix, W.D. 2002. Effects of the substrate on the determination of thin film mechanical properties by nanoindentation. Acta Materialia 50:23–28.
62. Nix, W.D. 1989. Mechanical properties of thin films. Metallurgical Transactions 20A:2217–2245.

63. Kahn, H., Heuer, A.H., and Ballarini, R. 2001. On-chip testing of mechanical properties of MEMS devices. MRS Bulletin 26:300–301.
64. Guckel, H., et al. 1992. Diagnostic microstructures for the measurement of instrinsic strain in thin films. Journal of Micromechanics and Microengineering:86–95.
65. Cho, H. S., et al. 2002. Tensile, creep and fatigue properties of LIGA nickel structures. IEEE. The 15th IEEE International Conference on MEMS, pp. 439–442.
66. Brown, S.B., Van Arsdell, W., and Muhlstein, C.L. 1997. Materials reliability in MEMS devices. International Conference on Solid-State Sensors and Actuators, pp. 591–593.
67. Connelly, T., McHugh, P.E., and Bruzzi, M. Dec 2005. A review of deformation and fatigue of metals at small size scales. Fatigue and Fracture of Engineering Materials and Structures 28:1119–1152.
68. Zhang, X., and Tang, W.C. 1995. Viscous air damping in laterally driven microresonators. Sensors and Materials 7:145.
69. Van Spengen, M., et al. 2003. A low frequency electrical test set-up for the reliability assessment of capacitive RF MEMS switches. Journal of Micromechanics and Microengineering 13:604–612.
70. Mastrangelo, C.H., and Hsu, C.H. 1992. Proceedings of the Solid State Sensor and Actuator Workshop. New York, NY: IEEE, pp. 208–212.
71. Bhushan, B. 2003. Adhesion and stiction: mechanisms, measurement techniques, and methods for reduction. Journal of Vacuum Science & Technology: Microelectronics and Nanometer Structures 21:262–2296.
72. Tanner, D. M., et al. 2005. Accelerating Aging Failures in MEMS Devices. San Jose, CA: IEEE. IEEE 43rd Annual International Reliability Physics Symposium, pp. 317–324.
73. Bazu, M., et al. 2007. Quantitative accelerated life testing of MEMS accelerometers. Sensors 7:2846–2859, 1424–8220.
74. Vudathu, S.P., and Laur, R. 2007. A Design Methodology for the Yield Enhancement of MEMS Designs with Respect to Process Induced Variations. Reno, NV: IEEE. 57th IEEE Electronics Components and Technology Conference (ECTC).
75. Nemeth, N.N., et al. 2003. Probabilistic Weibull behavior and mechanical properties of MEMS brittle materials. Journal of Material Science 38:4087–4113.
76. Allen, M., et al. 2004. Reliability-based analysis and design optimization of electrostatically actuated MEMS. Computers and Structures 82:1007–1020.
77. Espinosa, H.D., Prorok, B.C., and Fischer, M. 2003. Methodology for determining mechanical properties of freestanding thin films and MEMS materials. Journal of Mechanics and Physics of Solids 51:47–67.
78. Kayali, S. 1997. Methodology for MEMS Reliability Evaluation and Qualification. Pasadena, CA: MEMS Reliability and Qualification Workshop.

Subject Index

A
Acceleration factor, 4, 9, 19, 21–22, 30, 41, 132, 160, 167, 191–192, 232, 244, 278–279, 281
Acceleration models, 9, 19, 21, 24, 25, 29, 228, 279
Activation energy, 21–22, 30, 39, 125, 147, 190–191, 279, 281
Adaptive optics (AO), 221
Analog Devices accelerometer, 30–32, 99, 243–244
Anti-stiction coating, 100, 206, 208–210
Atomic Force Microscopy (AFM), 203–204
Auger analysis, 206–207
Automotive grades, 6, 19, 180, 228, 245, 248, 263
Automotive standards, 240–242, 281

B
Bathtub curve, 4, 10–12, 24, 193
Bimodal distribution, 18
Boston Micromachines Corporation, 188–189
Burn-in, 11, 64, 69, 192–193, 213

C
Classes of MEMS devices, 216–217
Constant acceleration, 104, 228, 231–232, 234, 236–237, 243, 248–249
Contamination, 5, 37, 52, 55, 64, 66–67, 69, 76–77, 98, 166, 169, 206, 211, 221, 243, 264, 279
Corrosion, 20, 23, 64–65, 70, 78, 80, 119, 148, 168–170, 182, 199, 230, 232, 247
Corrosion, galvanic, 64, 148, 159–170, 199, 231
Creep, 20, 22, 25, 29, 36–37, 41, 61, 73, 85, 105, 114–118, 122, 164, 171, 188, 190–192, 272, 277
Cumulative distribution function, 9, 12–13, 16

Curvature, 49, 56, 60, 117, 183–192, 208, 213, 222, 266, 272, 273–276

D
Deformable mirrors, 221–223
Design analysis, 49–50
Design for reliability (DfR), 6, 86, 182, 185–188, 254, 282–285
Dielectric charging, 6, 20, 34–37, 40, 51, 62, 118, 123–137, 154–159
Diurnal humidity variation, 237–239
Diurnal temperature variation, 19, 21, 228–230, 234–235, 238, 241–242, 244, 279
Drop test, 92, 241, 244
Dynamic Interferometry, 223

E
Elastic limit, 87, 99
Electron Beam Scatter Detector (EBSD), 199–200
Energy Dispersive X-ray Analysis (EDX, EDS, EDAX), 204–206
ESD, 6, 62–63, 85, 138, 144–146, 240, 242
ESD standards, *see* ESD
Exponential distribution, 4, 12–13

F
Failure analysis, 6, 70, 100, 179–213, 216, 283
Failure mechanisms, 4–6, 12, 19–20, 25, 41, 44, 54, 64, 71, 73, 77, 179–180, 190, 192, 212, 223, 230, 243, 247, 268, 270–271, 279–280, 282
Failure modes, 25, 34, 36, 43–81, 85–114, 123–148, 154, 179–180, 253–255, 262, 268, 278–279, 283–284
Failure rate, 6, 10–12, 15, 24, 31, 33, 51, 179, 192–193, 216, 244, 255, 278–279, 281
Fatigue, 5–6, 18, 20, 36–37, 48, 59–60, 65, 85, 105, 110, 118–122, 135, 182, 272, 276

A.L. Hartzell et al., *MEMS Reliability*, MEMS Reference Shelf, DOI 10.1007/978-1-4419-6018-4, © Springer Science+Business Media, LLC 2011

CPSIA information can be obtained at www.ICGtesting.com
Printed in the USA
LVOW010038121212

311237LV00008B/135/P

9 781441 960177